H. K. Porter

Light Locomotives

A Handbook

H. K. Porter

Light Locomotives
A Handbook

ISBN/EAN: 9783743372894

Manufactured in Europe, USA, Canada, Australia, Japa

Cover: Foto ©berggeist007 / pixelio.de

Manufactured and distributed by brebook publishing software (www.brebook.com)

H. K. Porter

Light Locomotives

LOCOMOTIVES.

H. K. PORTER & CO.

PITTSBURGH, PA.

SIXTH EDITION.

1889.

BUFFALO, N. Y.
MATTHEWS, NORTHRUP & CO., ART-PRINTING WORKS,
Office of the "Buffalo Morning Express."

H. K. PORTER & CO.,

BUILDERS OF LIGHT LOCOMOTIVES.

PITTSBURGH, PA.

OFFICE, Corner of Smithfield and Water Streets, in Monongahela House Building.

WORKS, On Allegheny Valley R. R., 49th to 50th Streets.

BUSINESS ESTABLISHED 1866.

SMITH & PORTER .	1866–1871
PORTER, BELL & CO. .	1871–1878
H. K. PORTER & CO.	1878

FOR INDEX SEE LAST PAGE.

LIGHT LOCOMOTIVES.

Our EXCLUSIVE SPECIALTY is the manufacture of Light Locomotives in every variety of size and style, and *for any practicable gauge of track*, to meet the requirements of many kinds of service for which ordinary locomotives are not practical or are not economical.

Our LOCATION in the city of Pittsburgh, Pa., affords us unusual advantages in obtaining supplies and shipping locomotives. Our shops were built by us, and stocked with tools especially adapted to our business. Our designs and methods of construction are not mere copies on a reduced scale from heavy locomotives, but are the results of our experience in this specialty for many years. *Natural gas* is used for forging and case-hardening. We use only the best materials. Our shop force is well drilled, most of the workmen having been educated in our employ, and all of them take pride in the good reputation of the shop.

Our DUPLICATE SYSTEM is a most valuable feature, to which we invite special attention. By means of original and duplicate drawings and records, and of standard gauges and templets, and of special tools and machines, each locomotive is made interchangeable with all others of the same size and class. This reduces the cost of repairs of our locomotives to the minimum and saves their owners from any expense for patterns or shops. A good engineer is competent to attach duplicate parts and usually without losing a trip. We furnish with every locomotive a LIST OF NAMES OF PARTS, to save mistakes in ordering supplies. *Our duplicate system differs* in one important item from that of other shops. *We always keep on hand*, independent and ahead of orders, a full stock of fitted duplicate parts for our standard designs and sizes, so that orders for repairs are filled immediately upon receipt. This practically insures our locomotives against loss of time, although customers in foreign countries or at a great distance may find it desirable to order with their locomotives a few extra parts most liable to wear or injury. Our records show that *90 per cent of orders for supplies are filled from fitted stock on hand*, 63 per cent being shipped on the day of receipt of order, and 27 per cent on the next day, because orders were received too near the close of business hours. Of the remainder, 5 per cent were shipped two days and 5 per cent more than two days after the receipt of order. This includes all shipments of supplies except departures from standard designs made by customers' instructions, and some parts differing with gauge of track which are not kept on hand for unusual gauges.

QUICK DELIVERY OF LOCOMOTIVES and prompt completion on or before contract time is secured by our system of construction. We can usually fill orders for locomotives inside of 60 or 90 days and occasionally in 30 days. We request correspondents not to ask for earlier delivery than necessary, as we have only limited facilities for storing engines.

IMMEDIATE DELIVERY OF LOCOMOTIVES is not often to be expected. But for a number of years we have endeavored to keep on hand completed locomotives of several sizes for wide and for narrow gauge, suitable for contractor's use, steel works, logging roads, suburban roads, etc. When any of these stock locomotives are sold, whether before or after completion, another one is at once put under construction. *We do not buy or sell second hand locomotives.*

OUR GUARANTY.

We guarantee all our locomotives to be according to specifications ; to be of best work and material, accurately constructed to our duplicate system ; to be efficient in service and to come up to their hauling capacity as given and explained in this catalogue.

We offer the very best work, of designs adapted to special requirements, accurate, interchangeable, and durable, at short notice and reasonable prices.

Our locomotives are in operation in nearly every State and Territory of United States, and in Canada, in the West Indies and Mexico, in different parts of South America, and in Japan, and we consider them our best advertisement, and their owners as our best references. On an average over half our orders are from old customers, and most of the rest are given from information received or from personal knowledge of the efficiency of our engines at work.

PRICES OF LOCOMOTIVES.

It is not practicable to name prices in this catalogue. On application of customers we will make propositions, with photographs and specifications for locomotives guaranteed to do the required work. Such applications should state

1.   The gauge of track, length of road, kind of fuel, weight of rail, and radius of sharpest curve.

2.   The steepest grade, with its length, for loaded cars to go up (also the same for empty cars if they return empty).

3.   The number of cars to be hauled in each train and the weight of each car and of its load.

4.   The total amount of freight to be carried one way daily.

When customers have previously determined on the size and style of locomotives they require, we would still request the above information, as we may be able to suggest some less expensive and more satisfactory design ; and also because we wish in all cases to be convinced ourselves that locomotives furnished by us are of such power and design as are best adapted to perform the work, and so will be of credit to us, and of the utmost benefit to their owners.

With orders for locomotives it is desirable that the following information be given promptly :

1.   The gauge of track (exact space in the clear between rails);

2.   The kind of fuel ;

3.   The height of the centre of the car couplings above the rail ;

4.   (At later convenience) the lettering for cab and tank.

THE STANDARD SPECIFICATIONS

of our LIGHT LOCOMOTIVES include axles, tires, guides, crank-pins, rods, links and springs of steel; valve gear and other working joints, links and blocks of case-hardened steel with extra long bearings, with hardened steel pins and thimbles; iron frames solidly forged; cylinders and all cast-iron wearing surfaces of close, hard charcoal mixture of metal; wearing brasses ingot copper and as large a proportion of tin as can be worked; all moveable nuts and bolts case-hardened; all parts drilled, planed, turned and fitted to gauges and templets, and interchangeable; all bolts of U. S. standard thread; all cocks to standard gas-taps; all material and workmanship of the very best; painting and finish neat and suited to the service throughout. Boiler of homogeneous cast steel plates; lap-welded flues, set with copper ferrules at the fire-box ends; all caulking done with a blunt tool on bevelled edges by the patent concave process; rivets hand riveted by the latest and best patent method; boiler tested before lagging to 180 lbs. hydraulic pressure, and engine fired up and worked by its own steam on friction rollers before shipment. Tank of steel plates.

Special attention is given to secure for all of our locomotives thorough fitness in all details for the service required; also compactness and accessibility of machinery, and convenience and perfect control of all working levers, gauges, etc., by the engineer.

Our locomotives are furnished with pump and injector (or two injectors and no pump), with seamless copper pipe connections; sand-box; bell (except mine locomotives, motors and some special styles); safety and relief valves, steam gauge, cab-lamp, cylinder oilers, blow-off, heater, blower, gauge, pet, sprinkling, and other cocks; tool-box and cushion; tools, including two screw-jacks, tallow and oil cans, spanner and flat wrenches to fit all bolts and nuts; monkey-wrench, steel and copper hammers; chisels, pinch-bar, poker, scraper, and torch.

Headlights, driver or power-brakes, syphon pumps, etc., are extra.

Unless otherwise agreed, our delivery is free on board cars at our shops. We can obtain advantageous freight rates to all accessible points. For foreign shipments we are prepared to include in our propositions the taking apart of locomotives, protecting from rust, boxing, and prepaying freight and lighterage charges to the vessel's dock.

The illustrations and descriptions herein presented comprise only our leading styles and sizes; we have many modifications of these, besides other special patterns and designs, and are also ready to prepare other designs for peculiar cases, or to build to required specifications.

EIGHT-WHEEL PASSENGER LOCOMOTIVE.

Cylinders ⎰ diameter........................	11 inches.	12 inches.	12 inches.	13 inches.
⎱ stroke	16 inches.	16 inches.	18 inches.	18 inches.
Diameter of driving wheels................	40 inches.	40 to 44 in.	44 to 48 in.	44 to 48 in.
Diameter of truck wheels	20 inches.	18 to 20 in.	20 to 22 in.	20 to 22 in.
Rigid wheel-base of engine.	6 ft. 0 in.	6 ft. 6 in	6 ft. 9 in.	6 ft. 9 in.
Total wheel-base of engine..........	15 ft. 6 in.	16 ft. 4 in.	16 ft.10 in.	17 ft. 7 in.
Wheel-base of engine and tender.........	32 ft. 4 in.	34 ft. 3 in.	34 ft. 9 in.	37 ft. 5 in.
Length over all of engine and tender.	39 ft. 9 in.	42 ft. 5 in.	42 ft.11 in.	46 ft. 2 in.
Weight of engine in working order..... ...	34,000 lb.	37,000 lb.	39,000 lb.	44,000 lb.
Weight on driving wheels.....	23,000 lb.	25,000 lb.	26,000 lb.	29,500 lb.
Weight on four-wheel truck....	11,000 lb.	12,000 lb.	13,000 lb.	14,500 lb.
Water capacity of tender tank.......... ..	1,050 gals.	1,200 gals.	1,200 gals.	1,400 gals.
Weight per yard of lightest steel rail advised	30 lb.	30 lb.	30 lb.	35 lb.
Hauling capacity on a level, in tons of 2,000 lb.............	600 tons.	650 tons.	700 tons.	800 tons.

To compute the hauling capacity on any practicable grade, refer to Table II., page 47.

The "Eight-wheel" or "American" pattern of locomotive is deservedly a favorite for general use on broad-gauge roads throughout the United States, and hence has been very largely adopted by narrow-gauge roads.

We believe, however, that a narrow-gauge engine, or a light engine for wide gauge, should be something more than a miniature copy of a full size standard-gauge engine, and that the construction necessary on a large engine should be simplified on a small engine where it can be done advantageously.

We regard the "Eight-wheel" pattern, especially the smaller sizes, as less desirable than some other designs in the following particulars :

The weight is not distributed to secure the maximum of power, the proportion of dead to useful weight being necessarily very large.

The truck wheels are necessarily of smaller diameter than is advisable for high speeds ; or to secure larger truck wheels the boiler is set higher, and the centre of weight raised more than is desirable for fast running.

While we recommend the design illustrated on page 6 in preference to the "Eight-wheel" pattern, we wish to meet the views of all customers, and are prepared to furnish this style of sizes as specified.

NOTE.—Refer to page 46 for explanation of hauling capacity; for regular work locomotives should be used at one-half or two-thirds of their full capacity or at a less proportion for fast speeds.

For actual performances see WORKING REPORTS on pages 90 and 91.

SIX-WHEEL PASSENGER LOCOMOTIVE.

Cylinders { diameter	11 inches.	12 inches.	12 inches.	13 inches.
Cylinders { stroke	16 inches.	16 inches.	18 inches.	18 inches.
Diameter of driving wheels	40 inches.	40 to 44 in.	44 to 48 in.	44 to 48 in.
Diameter of truck wheels	26 inches.	26 to 30 in.	30 inches.	30 inches.
Rigid wheel-base of engine	6 ft. 0 in.	6 ft. 6 in.	6 ft. 9 in.	6 ft. 9 in.
Total wheel-base of engine	16 ft. 2 in.	16 ft. 10 in.	17 ft. 4 in.	18 ft. 1 in.
Wheel-base of engine and tender	32 ft. 6 in.	34 ft. 10 in.	35 ft. 4 in.	37 ft. 11 in.
Length over all of engine and tender	39 ft. 0 in.	43 ft. 5 in.	43 ft. 11 in.	46 ft. 6 in.
Weight of engine in working order	33,000 lb.	36,000 lb.	38,000 lb.	43,000 lb.
Weight on driving wheels	25,000 lb.	27,000 lb.	28,500 lb.	32,500 lb.
Weight on two-wheel radial-bar truck	8,000 lb.	9,000 lb.	9,500 lb.	10,500 lb.
Water capacity of tender tank	1,050 gals.	1,200 gals.	1,200 gals.	1,400 gals.
Weight per yard of lightest steel rail advised	30 lb.	30 lb.	30 lb.	35 lb.
Hauling capacity on a level, in tons of 2,000 lb.	650 tons.	700 tons.	750 tons.	850 tons.

To compute the hauling capacity on any practicable grade, refer to Table II., page 47.

The pattern of locomotive illustrated on the opposite page was designed by us for *fast passenger service and long runs* on narrow gauge, and also for light work on standard gauge, and has proved extremely powerful and fast. The special advantages of this pattern over others for such service are :

Economical distribution of weight, securing the greatest proportion of useful weight, and consequently the greatest power, as well as ease on track.

The centre of weight is extremely low, securing unusual stability; and the pony truck wheels are of large diameter, rendering the engine capable of very high speed with perfect safety.

The unusually long flexible wheel-base secures great ease of motion, even on a rough track ; and the short rigid wheel-base and superior curving qualities of the truck permit passing sharp curves even at a high speed.

The truck axle and machinery are proportioned to the load to be upheld, and better able to endure severe shocks than the smaller axles and lighter machinery of the four-wheel truck. At the same time simplicity is attained and useless gear avoided. Curves of a 150 feet radius, speed of 40 to 60 miles per hour, and runs of 150 to 200 miles per day are practicable.

The same general style, with smaller drivers, and of sufficient weight to utilize them, is very efficient for freight or for mixed traffic, or for passenger service on heavy grades, and is by many preferred to the "Mogul" style (page 16).

NOTE.—Refer to page 46 for explanation of hauling capacity; for regular work locomotives should be used at one-half or two-thirds of their full capacity or at a less proportion for fast speeds.

For actual performances, see WORKING REPORTS on pages 89 to 91.

MEDIUM PASSENGER LOCOMOTIVE.

These engines are designed for passenger or mixed service, for shorter runs and slower speed than the patterns shown on pages 4 and 6.

They will readily pass curves of 125 feet radius, and a speed of 30 to 40 miles per hour is attainable under favorable conditions.

The very large proportion of weight on the driving wheels adapts these locomotives for steep grades, for heavy loads and for quick stopping and starting of trains. In most cases they are practically as efficient as the next larger sizes of the styles on pages 4 and 6.

Cylinders { diameter		10 inches.	11 inches.
{ stroke		16 inches.	16 inches.
Diameter of driving wheels		36 to 40 in.	40 to 44 in.
Diameter of truck wheels		24 to 26 in.	26 to 28 in.
Rigid wheel-base of engine		6 ft. 6 in.	6 ft. 6 in.
Total wheel-base of engine		13 ft. 3 in.	14 ft. 3 in.
Wheel-base of engine and tender		29 ft. 6 in.	32 ft. 9 in.
Length over all of engine and tender		36 ft. 6 in.	40 ft. 0 in.
Weight of engine in working order		28,000 lb.	32,000 lb.
Weight on driving wheels		24,000 lb.	26,000 lb.
Weight on two-wheel radial-bar truck		4,000 lb.	6,000 lb.
Water capacity of tender tank		800 gals.	1,050 gals.
Weight per yard of lightest steel rail advised		30 lb.	30 lb.

Hauling capacity on a level, in tons of 2,000 lb.	625 tons.	700 tons.

To compute the hauling capacity on any practicable grade, refer to Table II., page 47.

NOTE.—Refer to page 46 for explanation of hauling capacity; for regular work locomotives should be used at one-half or two-thirds of their full capacity, or at a less proportion for fast speeds.

For actual performances, see WORKING REPORTS on pages 87, 88, 89 and 144.

FOUR-WHEEL-CONNECTED SADDLE-TANK LOCOMOTIVE, WITH FRONT TRUCK.

These engines are well adapted for suburban roads where the grades and loads are heavy, and where the run is not long enough to require a tender tank. As the weight of the water is used for traction, and there is no tender, these engines can haul larger trains than those shown on the opposite page. The relative advantage increases with the grade.

In most cases the "Back-Truck," design described on page 19 or page 21 is preferable, as it admits more fuel space and more cab room.

The engines may be run without turning, and are adapted to either wide or narrow gauge.

Cylinders { diameter	10 inches.	11 inches.
Cylinders { stroke	16 inches.	16 inches.
Diameter of driving wheels	33 to 40 in.	36 to 40 in.
Diameter of truck wheels	22 to 26 in.	24 to 26 in.
Rigid wheel-base	6 ft. 6 in.	6 ft. 6 in.
Total wheel-base	13 ft. 3 in.	14 ft. 3 in.
Length over all	26 ft. 9 in.	28 ft. 0 in.
Weight in working order	33,000 lb.	37,000 lb.
Weight on driving wheels	27,000 lb.	30,000 lb.
Weight on two-wheel radial-bar truck	6,000 lb.	7,000 lb.
Water capacity of saddle tank	500 gals.	600 gals.
Weight per yard of lightest steel rail advised	30 lb.	35 lb.
Hauling capacity on a level, in tons of 2,000 lb.	675 tons.	800 tons.

To compute the hauling capacity on any practicable grade, refer to Table I, page 47.

NOTE.—Refer to page 46 for explanation of hauling capacity; for regular work locomotives should be used at one-half or two-thirds of their full capacity, or at a less proportion for fast speeds.

LIGHT PASSENGER LOCOMOTIVE.

These engines are designed for light work on light rails. They will pass curves of 75 feet radius; and are capable of a speed of 25 to 35 miles per hour.

We are prepared to build smaller engines of this style.

Cylinders { diameter.....	8 inches.	9 inches.
{ stroke..........	14 inches.	14 inches.
Diameter of driving wheels........	30 inches.	36 inches.
Diameter of truck wheels..........	18 inches.	22 inches.
Rigid wheel-base of engine....	5 ft. 0 in.	5 ft. 9 in.
Total wheel-base of engine..........	9 ft. 0 in.	10 ft. 9 in.
Wheel-base of engine and tender....	23 ft. 0 in.	25 ft. 6 in.
Length over all of engine and tender..........	30 ft. 0 in.	32 ft. 6 in.
Weight of engine in working order..........	16,000 lb.	20,000 lb.
Weight on driving wheels..........	13,500 lb.	17,000 lb.
Weight on two-wheel radial-bar truck..........	2,500 lb.	3,000 lb.
Water capacity of tender tank	500 gals.	500 gals.
Weight per yard of lightest steel rail advised..........	20 lb.	25 lb.
Hauling capacity on a level, in tons of 2,000 lb.	350 tons.	475 tons.

To compute the hauling capacity on any practicable grade, refer to Table II., page 47.

NOTE.—Refer to page 46 for explanation of hauling capacity; for regular work locomotives should be used at one-half or two-thirds of their full capacity, or at a less proportion for fast speeds.

For actual performances, see WORKING REPORTS, on pages 86, 87, 138 and 140.

LIGHT FOUR-WHEEL-CONNECTED SADDLE-TANK LOCOMOTIVE, WITH FRONT TRUCK.

These engines are adapted for light suburban traffic and other service where a greater speed is needed than is easily attainable by four-wheel-connected tank locomotives, and where the run is not long enough to require a tender. As the weight of the water is used for traction, and there is no tender, these engines can haul heavier trains than those shown on the opposite page. This relative advantage increases with the grade.

In most cases the "Back-Truck" design, described on pages 19, 20 or 21 is preferable, as it admits more fuel space and more cab room.

These engines may be run without turning, and are adapted to either wide or narrow gauge.

Cylinders { diameter	8 inches.	9 inches.
Cylinders { stroke	14 inches.	14 inches.
Diameter of driving wheels	30 inches.	33 inches.
Diameter of truck wheels	18 inches.	22 inches.
Rigid wheel-base	5 ft. 0 in.	5 ft. 9 in.
Total wheel-base	8 ft. 7 in.	10 ft. 9 in.
Length over all	17 ft. 6 in.	19 ft. 9 in.
Weight in working order	21,500 lb.	25,000 lb.
Weight on driving wheels	17,000 lb.	21,000 lb.
Weight on two-wheel radial-bar truck	3,500 lb.	4,000 lb.
Water capacity of saddle tank	275 gals.	325 gals.
Weight per yard of lightest steel rail advised	20 lb.	25 lb.

Hauling capacity on a level, in tons of 2,000 lb.	425 tons.	550 tons.

To compute the hauling capacity on any practicable grade, refer to Table I., page 47.

NOTE.—Refer to page 46 for explanation of hauling capacity; for regular work locomotives should be used at one-half or two-thirds of their full capacity.

SIX-WHEEL-CONNECTED LOCOMOTIVE, WITH TENDER.

Cylinders	diameter	10 inches.	11 inches.	12 inches.	12 inches.	13 inches.
	stroke	16 inches.	16 inches.	16 inches.	18 inches.	18 inches.
Diameter of driving wheels.. ...		33 inches.	33 inches.	36 inches.	36 inches.	40 inches.
Wheel-base of engine...........		7 ft. 8 in.	8 ft. 1 in.	8 ft. 1 in.	9 ft. 0 in.	10 ft. 0 in.
Wheel-base of engine and tender.		28 ft. 0 in.	28 ft. 0 in.	29 ft. 0 in.	29 ft. 6 in.	30 ft. 0 in.
Length over all of engine and tender.......................		35 ft. 0 in.	39 ft. 0 in	40 ft. 0 in.	41 ft. 0 in.	41 ft. 6 in.
Weight of engine in working order (all on drivers).............		28,000 lb.	30,000 lb.	33,000 lb.	36,000 lb.	41,000 lb.
Water capacity of tender tank ..		800 gals.	1,050 gals.	1,050 gals.	1,050 gals.	1,200 gals.
Weight per yard of lightest steel rail advised		25 lb.	30 lb.	30 lb.	30 lb.	35 lb.
Hauling capacity on a level, in tons of 2,000 lb.		750 tons.	800 tons.	875 tons.	975 tons.	1,100 tons.

To compute the hauling capacity on any practicable grade, refer to Table II., page 47.

For SADDLE-TANK LOCOMOTIVES of this class, see page 23.

These engines are equalized between rear and centre drivers ; they also have a cross equalizer at front drivers. The centre drivers are without flanges. The engines are easy on the track, and curve well up to a speed of 15 to 20 miles per hour. Having all their weight on drivers, and having a short wheel base, they are specially adapted to hauling heavy loads on steep grades and short curves, and in many cases are preferable to the "Mogul" described on page 16.

NOTE.—Refer to page 46 for explanation of hauling capacity; for regular work locomotives should be used at one-half or two-thirds of their full capacity.

For actual performances, see WORKING REPORTS on pages 98, 100, 101, 146 and 147.

LIGHT SIX-WHEEL-CONNECTED LOCOMOTIVE, WITH TENDER.

Cylinders		8 inches.	9 inches.	9½ in.
	diameter.............................	8 inches.	9 inches.	9½ in.
	stroke............................	14 inches.	14 inches.	14 inches.
Diameter of driving wheels....................		26 inches.	28 inches.	33 inches.
Wheel-base of engine............................		5 ft. 5 in.	5 ft. 10 in.	7 ft. 3 in.
Wheel-base of engine and tender...............		20 ft. 0 in.	21 ft. 0 in.	22 ft. 0 in.
Length over all of engine and tender...........		27 ft. 0 in.	28 ft. 0 in.	30 ft. 0 in.
Weight of engine in working order (all on drivers)...................................		16,000 lb.	18,500 lb.	22,000 lb.
Water capacity of tender tank..................		300 gals.	500 gals.	800 gals.
Weight per yard of lightest steel rail advised..		16 lb.	20 lb.	25 lb.
Hauling capacity on a level, in tons of 2,000 lb..		400 tons.	500 tons.	600 tons.

To compute the hauling capacity on any practicable grade, refer to Table II., page 47.

NOTE.—The 8 x 14 cylinders locomotive has four-wheeled tender.

For SADDLE TANK LOCOMOTIVES of this class, see page 22.

These engines are designed for local freight or mixed trains on light equipped roads narrow or standard gauge ; also for construction, and for special service where the run is longer than is expedient for saddle-tank engines. Curves of less than 100 feet radius are admissible. The centre drivers are without flanges. The weight on drivers is equalized in the same manner as the engines on page 12. We would advise that the running time should not exceed 15 miles per hour, although on easy grades and curves this style has run 30 miles per hour.

We are prepared to build smaller sizes of this style, and also to add a two-wheel pony truck (like page 16) ; but in most cases some other style would be preferable.

NOTE.—Refer to page 46 for explanation of hauling capacity ; for regular work locomotives should be used at one-half or two-thirds of their full capacity.

For actual performances, see WORKING REPORTS on pages 96, 97, 142 and 143.

Cylinders { diameter..........	11 inches.	12 inches.	12 inches.	13 inches.	14 inches.
Cylinders { stroke...............	16 inches.	16 inches.	18 inches.	18 inches.	20 inches.
Diameter of driving wheels.....	36 inches.	36 inches.	40 inches.	40 inches.	44 inches.
Diameter of truck wheels.	24 inches.	24 inches.	26 inches.	26 inches.	28 inches.
Rigid wheel-base of engine......	9 ft. 0 in.	9 ft. 0 in.	9 ft. 3 in.	11 ft. 5 in.	12 ft. 2 in.
Total wheel-base of engine......	14 ft. 6 in.	14 ft. 6 in.	15 ft. 0 in.	17 ft. 6 in.	18 ft. 2 in.
Wheel-base of engine and tender	33 ft. 0 in.	33 ft. 6 in.	35 ft. 2 in.	37 ft. 2 in.	38 ft. 0 in.
Length over all of engine and tender........................	40 ft. 6 in.	41 ft. 2 in.	42 ft. 8 in.	45 ft. 0 in.	45 ft. 8 in.
Weight of engine in working order	32,000 lb.	35,000 lb.	38,000 lb.	44,000 lb.	51,000 lb.
Weight on driving wheels..	27,500 lb.	30,500 lb.	33,000 lb.	38,000 lb.	43,000 lb.
Weight on two-wheel radial-bar truck	4,500 lb.	4,500 lb.	5,000 lb.	6,000 lb.	8,000 lb.
Water capacity of tender tank..	1,050 gals.	1,050 gals.	1,200 gals.	1,400 gals.	1,600 gals.
Weight per yard of lightest steel rail advised............	30 lb.	30 lb.	30 lb.	35 lb.	40 lb.
Hauling capacity on a level, in tons of 2,000 lb.	700 tons.	800 tons.	900 tons.	1,000 tons.	1,150 tons.

To compute the hauling capacity on any practicable grade, refer to Table II., page 47.

These engines are specially adapted for hauling freight on long roads where considerable speed is desired. They are also useful in hauling mixed trains or passenger trains on heavy grades.

Curves of 150 feet radius, a speed of 25 miles per hour, and daily mileage of 150 or more miles are practicable.

The rear and centre pairs of drivers, also the front drivers and the truck, are equalized together. The centre drivers are without flanges.

Our "Mogul" locomotives, by reason of their short rigid wheel-base and superior design of truck, are able to pass very sharp curves with ease. Their centre of weight is very low, which gives unusual stability and safety at high speed.

We are prepared to build smaller sizes of "Mogul" locomotives than are described on the opposite page.

NOTE.—Refer to page 46 for explanation of hauling capacity; for regular work locomotives should be used at one-half or two-thirds of their full capacity, or at a less proportion for fast speeds.

For actual performances, see WORKING REPORTS on pages 97 to 101, and 141 to 147.

"DOUBLE-ENDER" LOCOMOTIVE.

This style is especially adapted for suburban passenger roads of wide or narrow gauge, where a compact, fast engine is desired, which, by running equally well forward or back, requires no turn-table or Y. Sharp curves are admissible. On easy grades and straight track these engines are capable of a speed of 30 to 40 miles per hour. These engines are not intended for very heavy loads or excessive grades. Their motion is very easy, as both pairs of driving wheels are equalized and the weight is well distributed.

This style is adaped to narrow or wide gauges, and we are prepared to build several other sizes in addition to those given below.

The styles illustrated on pages 19, 20, 21, 86, and 37 may be preferable where heavy grades are to be overcome, or heavy trains are to be hauled.

Cylinders { diameter	8 inches	9 inches.	10 inches	12 inches.	14 inches.
{ stroke	14 inches.	14 inches.	16 inches.	18 inches.	20 inches.
Diameter of driving wheels	30 to 33 in.	33 to 36 in.	40 to 44 in.	44 to 48 in.	48 inches.
Diameter of truck wheels	16 to 18 in.	18 to 20 in.	22 to 24 in.	24 to 26 in.	26 inches.
Rigid wheel-base	5 ft. 0 in.	5 ft. 9 in.	6 ft. 6 in.	6 ft. 9 in.	7 ft. 0 in.
Total wheel-base	15 ft. 0 in.	15 ft. 9 in.	18 ft. 6 in.	20 ft. 0 in.	21 ft. 0 in.
Length over all	22 ft. 0 in.	24 ft. 0 in.	30 ft. 0 in.	32 ft. 6 in.	35 ft. 0 in.
Weight in working order	23,000 lb.	29,000 lb.	39,000 lb.	49,000 lb.	58,000 lb.
Weight on driving wheels	15,000 lb.	19,000 lb.	27,000 lb.	33,000 lb.	40,000 lb.
Weight on two trucks	8,000 lb.	10,000 lb.	12,000 lb.	16,000 lb.	18,000 lb.
Capacity of saddle tank	250 gals.	325 gals.	500 gals.	750 gals.	900 gals.
Weight per yard of lightest steel rail advised	20 lb.	25 lb.	30 lb.	35 lb.	40 lb.
Hauling capacity on a level, in tons of 2,000 lb.	350 tons.	475 tons.	650 tons.	850 tons.	1,000 tons.

To compute the hauling capacity on any practicable grade, refer to Table I., page 47.

NOTE.—The 8 x 14 and 9 x 14 cylinders are placed slightly inclined.

NOTE.—Refer to page 46 for explanation of hauling capacity; for regular work locomotives should be used at one-half or two-thirds of their full capacity, or at a less proportion for fast speeds.

For actual performances, see WORKING REPORTS on pages 86, 87 and 91.

"BACK TRUCK" LOCOMOTIVE.

(AS ADAPTED TO LOCAL PASSENGER SERVICE.)

This style is advisable for suburban roads, for passenger or mixed service, for either narrow or wide gauge, where considerable power combined with fast speed is required. No turn-table is needed, and the motion is easy both when running with the truck ahead or following. Very sharp curves are practicable. Speeds of 15 to 25 miles per hour on curves and grades, and 30 to 40 miles per hour under favorable circumstances may be attained. The driving wheels are equalized, the weight is well distributed ; and as a much larger proportion of the weight is used for traction, this style is usually preferable to the "double ender" style described on the opposite page.

We are prepared to build this general style of the smaller sizes described on page 20, but these are only suitable for very light work.

Cylinders { diameter ..	9 inches.	9½ inches	10 inches.	12 inches.	14 inches.	14 inches.
Cylinders { stroke	14 inches.	14 inches.	16 inches.	18 inches.	20 inches	24 inches.
Diameter of driving wheels...............	33 to 36 in.	33 to 36 in.	36 inches.	40 inches.	44 inches.	44 inches.
Diam. of truck wheels.	20 to 22 in.	20 to 22 in.	22 inches.	24 inches.	26 inches.	26 inches.
Rigid wheel-base......	4 ft. 6 in.	4 ft. 6 in.	5 ft. 3 in.	5 ft. 9 in.	6 ft. 3 in.	7 ft. 0 in.
Total wheel-base......	12 ft. 4 in.	12 ft. 6 in.	13 ft. 4 in.	14 ft. 0 in.	15 ft. 0 in.	15 ft. 9 in.
Length over all, including pilots	28 ft. 0 in.	29 ft. 0 in.	30 ft. 0 in.	31 ft. 0 in.	32 ft. 0 in.	34 ft. 0 in.
Weight in working order	28,000 lb.	31,000 lb.	35,000 lb.	44,000 lb.	54,000 lb.	59,000 lb.
Weight on driving wheels...............	21,000 lb.	24,000 lb.	27,500 lb.	35,500 lb.	45,000 lb.	50,000 lb.
Weight on two-wheel radial-bar truck.....	7,000 lb.	7,000 lb.	7,500 lb.	8,500 lb.	9,000 lb.	9,000 lb.
Water capacity of saddle-tank	375 gals.	400 gals.	500 gals.	750 gals.	900 gals.	1,000 gals.
Weight per yard of lightest steel rail advised............. ...	25 lb.	25 lb	30 lb.	35 lb.	45 lb.	50 lb.
Hauling capacity on a level, in tons of 2,000 lb.	525 tons.	625 tons.	725 tons.	925 tons.	1150 tons.	1300 tons.

To compute the hauling capacity on any practicable grade, refer to Table I., page 47.

NOTE.—The 9x14 and 9½x14 cylinders are placed slightly inclined.

NOTE.—Refer to page 46 for explanation of hauling capacity ; for regular work locomotives should be used at one-half or two-thirds of their full capacity, or at a less proportion for fast speeds.

For actual performances, see WORKING REPORTS on pages 88, 90, 91, 96 and 99.

LIGHT "BACK-TRUCK" LOCOMOTIVE.

(FOR LOGGING RAILROADS AND SIMILAR SERVICE.)

The style of locomotives illustrated and described below and on the opposite page is adapted to a great variety of service, including logging roads and plantation roads, where the track is uneven and the speed slow ; for switching and shifting where heavy loads are to be stopped and started promptly; and for local passenger traffic where the speed is fast and frequent stops are made.

For logging railroads and for plantations an open sheet-iron canopy is often used instead of a wooden cab, as shown on page 39.

These locomotives to a great extent combine the advantages and avoid the disadvantages of the "Double Ender" style on page 18, and of the "Four-Wheel-Connected" style on pages 24 and 26.

The driving wheels are equalized, and a very large (Continued on opposite page)

This cut shows cab with side sliding doors and bunker in rear part of cab (filled from outside if for coal) for cold climate.	This cut shows cab with open entrances at sides and sep rear fuel bunker for coal or wood, for warm climate.

Cylinders { diameter	6 inches.	7 inches.	8 inches.
{ stroke	10 inches.	12 inches.	14 inches.
Diameter of driving wheels	24 inches.	28 inches.	30 inches.
Diameter of truck wheels	16 inches.	16 inches.	18 inches.
Rigid wheel-base	4 ft. 0 in.	4 ft. 8 in.	5 ft. 0 in.
Total wheel-base	8 ft. 6 in.	9 ft. 1 in.	9 ft. 10 in.
Length over all	14 ft. 0 in.	16 ft. 4 in.	17 ft. 4 in.
Weight in working order	15,000 lb.	18,500 lb.	22,500 lb.
Weight on driving wheels	11,000 lb.	14,000 lb.	16,500 lb.
Weight on two-wheel radial-bar truck	4,000 lb.	4,500 lb.	5,000 lb.
Capacity of saddle tank	150 gals.	200 gals.	250 gals.
Weight per yard of lightest steel rail advised	16 lb.	16 lb.	20 lb.

Hauling capacity on a level, in tons of 2,000 lb	250 tons.	350 tons.	425 tons.

To compute the hauling capacity on any practicable grade, refer to Table I., page 47.

NOTE.—Refer to page 46 for explanation of hauling capacity; for regular work locomotives should be used at one-half or two-thirds of their full capacity.

For actual performances, see WORKING REPORTS on page 86 and pages 133 to 139.

"BACK TRUCK" LOCOMOTIVE.

(FOR LOGGING RAILROADS AND SIMILAR SERVICE.)

(*Continued from opposite page*) proportion of the weight is useful for traction. No turn-tables or Y's are required for these locomotives, since they run forward or backward with equal ease. They are adapted for sharp curves, and have an easy motion on rough track. They are capable of hauling large loads and attaining high rates of speed. They are very compact, and do not require heavy rails.

Other styles of four-driver tank engines with trucks are described on pages 9 and 11, and on pages 39, 35, 36 and 37.

This cut shows a cab with fuel bunker in back part of cab (filled from outside if for coal fuel), with side sliding doors adapted to Northern climate. We are prepared to build with a short cab and rear bunker for Southern climate, like the cut on page 19. Coal bunkers may also be arranged like page 24.

Cylinders { diameter...	9 inches.	9½ inches	10 inches.	12 inches.	14 inches.	14 inches.
Cylinders { stroke	14 inches.	14 inches.	16 inches.	18 inches.	20 inches.	24 inches.
Diameter of driving wheels.............	33 to 36 in.	33 to 36 in.	36 inches.	40 inches	44 inches.	44 inches.
Diam. of truck wheels.	20 to 22 in.	20 to 22 in.	22 inches.	24 inches.	26 inches.	26 inches.
Rigid wheel-base	4 ft. 6 in.	4 ft. 6 in.	5 ft. 3 in.	5 ft. 9 in.	6 ft. 3 in.	7 ft. 0 in.
Total wheel-base......	12 ft. 4 in.	12 ft. 10 in.	13 ft. 4 in.	14 ft.10 in.	15 ft. 6 in.	15 ft. 9 in.
Length over all, not including pilot.........	20 ft. 0 in.	21 ft. 0 in.	22 ft. 0 in.	23 ft. 0 in.	24 ft. 0 in.	26 ft. 0 in.
Weight in working order....	28,000 lb.	31,000 lb.	35,000 lb.	44,000 lb.	54,000 lb.	59,000 lb.
Weight on driving wheels......	21,000 lb.	24,000 lb.	27,500 lb.	35,500 lb.	45,000 lb.	50,000 lb.
Weight on two-wheel radial-bar truck	7,000 lb.	7,000 lb.	7,500 lb.	8,500 lb.	9,000 lb.	9,000 lb.
Water capacity of saddle tank............	375 gals.	400 gals.	500 gals.	700 gals.	900 gals.	1,000 gals.
Weight per yard of lightest steel rail advised................	25 lb.	25 lb.	30 lb.	35 lb.	45 lb.	45 lb.
Hauling capacity on a level, in tons of 2,000 lb.	525 tons.	625 tons.	725 tons.	925 tons.	1,150 tons.	1,300 tons.

To compute the hauling capacity on any practicable grade, refer to Table I., page 47.

NOTE.—Refer to page 46 for explanation of hauling capacity; for regular work locomotives should be used at one-half or two-thirds of their full capacity.

Note.—The 9 x 14 and 9½ x 14 cylinders are placed slightly inclined.

For actual performances, see WORKING REPORTS on pages 97, 100, 101, 108, 125 and pages 140 to 147.

LIGHT SIX-WHEEL-CONNECTED TANK LOCOMOTIVE.

These locomotives are specially useful on short runs where considerable loads are to be taken up steep grades. They are well adapted for switching and other special service.

The 9 x 14 and the 9½ x 14 sizes are good engines for suburban roads with steep grades, where engines with separate tenders are not desired ; for such service we often build with rear fuel bunker, two pilots, and two headlight brackets. For switching and other similar service, the fuel bunker of our six-wheel-connected tank engines may be at the rear, or inside the cab, as in our four-wheel-connected tank engines on page 26, or at the sides, as shown on page 24. For (*Continued on opposite page*)

Cylinders	diameter	... 8 inches.	9 inches.	9½ inches
	stroke	14 inches.	14 inches.	14 inches.
Diameter of driving wheels		26 inches.	28 inches.	30 inches.
Wheel-base		5 ft. 5 in.	5 ft. 10 in.	7 ft. 3 in.
Length over all		16 ft. 0 in.	17 ft. 0 in.	18 ft. 0 in.
Weight in working order (all on drivers)		20,000 lb.	25,000 lb.	28,000 lb.
Capacity of saddle tank		250 gals.	325 gals.	400 gals.
Weight per yard of lightest steel rail advised		20 lb.	25 lb.	25 lb.

Hauling capacity on a level, in tons
of 2,000 lb. .... 500 tons. | 625 tons. | 700 tons.

To compute the hauling capacity on any practicable grade, refer to Table I., page 47.

Six-wheel-connected locomotives, with tender instead of saddle tank, are shown on page 14.

NOTE.—Refer to page 46 for explanation of hauling capacity ; for regular work locomotives should be used at one-half or two-thirds of their full capacity.

For actual performances, see WORKING REPORTS on pages 102, 103, 104, 140, 141 and 142.

SIX-WHEEL-CONNECTED TANK LOCOMOTIVE.

(*Continued from opposite page*) logging railroads and freight work the cabs may be built with side sliding doors.

The 14 x 20 locomotive is specially adapted to wide gauge; the other sizes may be built to wide or narrow gauge.

The short wheel bases of these engines allow them to pass sharp curves, and our method of equalizing weight on drivers makes them easy in their motion. They may also be built with pony truck at the rear end, but there is seldom any advantage in this construction.

Cylinders	diameter	10 inches.	12 inches.	12 inches.	14 inches.
	stroke	16 inches.	16 inches.	18 inches.	20 inches.
Diameter of driving wheels		30 inches.	33 inches.	36 inches.	40 inches.
Wheel-base		7 ft. 8 in.	8 ft. 1 in.	9 ft. 0 in.	10 ft. 0 in.
Length over all		19 ft. 0 in.	20 ft. 0 in.	20 ft. 6 in.	21 ft. 6 in.
Weight in working order (all on drivers)		33,000 lb.	38,000 lb.	43,000 lb.	51,000 lb.
Capacity of saddle tank		600 gals.	750 gals.	750 gals.	900 gals.
Weight per yard of lightest steel rail advised		30 lb.	35 lb.	35 lb.	45 lb.
Hauling capacity on a level, in tons of 2,000 lb.		850 tons.	950 tons.	1150 tons.	1350 tons.

To compute the hauling capacity on any practicable grade, refer to Table I., page 47.

Six-wheel-connected locomotives, with tender instead of saddle tank, are shown on page 12.

NOTE.—Refer to page 46 for explanation of hauling capacity; for regular work locomotives should be used at one-half or two-thirds of their full capacity.

For actual performances, see WORKING REPORTS on pages 104, 105, 145 and 147.

FOUR-WHEEL-CONNECTED TANK LOCOMOTIVE.

These engines are designed for shifting and special service.

The 9½ x 14 and 10 x 16 are adapted to either wide or narrow gauge, but the larger sizes are not advisable for much narrower than 56½ inches gauge.

These engines have a cross equalizer at front drivers, by which to a considerable extent the uneven motion of an ordinary four-wheel engine is avoided. They are adapted to sharp curves, steep grades, slow speed, and heavy loads.

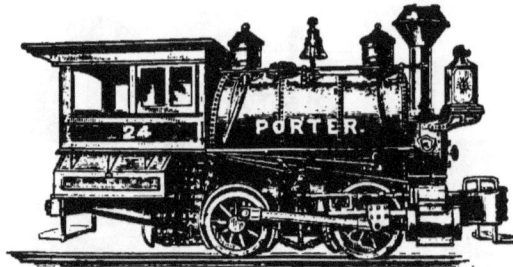

Cylinders {	diameter	9½ inches	10 inches.	12 inches.	14 inches.	14 inches.
	stroke	14 inches.	16 inches.	18 inches.	20 inches.	24 inches.
Diameter of driving wheels.		30 inches.	33 inches.	36 inches.	40 inches.	44 inches.
Wheel-base		4 ft. 6 in.	5 ft. 3 in.	5 ft. 9 in.	6 ft. 3 in.	7 ft. 0 in.
Length over all		16 ft. 9 in.	17 ft. 2 in.	19 ft. 9 in.	20 ft. 0 in.	22 ft. 0 in.
Weight in working order (all on drivers)		25,000 lb.	29,000 lb.	39,000 lb.	48,000 lb.	53,000 lb.
Capacity of saddle tank		400 gals.	500 gals.	750 gals.	900 gals.	900 gals.
Weight per yard of lightest steel rail advised..................		30 lb.	35 lb.	40 lb.	50 lb.	50 lb.
Hauling capacity on a level, in tons of 2,000 lb.........................		650 tons.	750 tons.	1,000 tons.	1,250 tons.	1,400 tons.

To compute the hauling capacity on any practicable grade, refer to Table I., page 47.

NOTE.—Refer to page 46 for explanation of hauling capacity; for regular work locomotives should be used at one-half or two-thirds of their full capacity

For actual performances, see WORKING REPORTS on pages 106, 107, 108, 109 and 145.

NOTE.—The 9½ x 14 cylinders are placed slightly inclined.

FOUR-WHEEL-CONNECTED LOCOMOTIVE, WITH TENDER.

These locomotives are designed for the same general service as those on the opposite page, but where the rail requires a lighter engine, or the length of the run makes a separate tender desirable. A four-wheel tender is usually sufficient, but we are prepared to build with eight-wheel tender if desired.

A four-wheel locomotive must necessarily have an uneven motion, for a perfect equalization of the weight is impossible. Except for slow speed and short runs, the locomotives shown on pages 19, 21, 37, and 36 are preferable to the four-wheel connected style.

Cylinders	diameter	9½ inches	10 inches.	12 inches.	14 inches.	14 inches.
	stroke	14 inches.	16 inches.	18 inches.	20 inches.	24 inches.
Diameter of driving wheels		33 inches.	36 inches.	40 inches.	44 inches.	48 inches.
Wheel-base of engine		4 ft. 6 in.	5 ft. 3 in.	5 ft. 9 in.	6 ft. 3 in.	7 ft. 0 in.
Wheel-base of engine and tender		20 ft. 0 in.	21 ft. 0 in.	22 ft. 6 in.	24 ft. 6 in.	26 ft. 0 in.
Length over all of engine and tender		28 ft. 0 in.	29 ft. 6 in.	31 ft. 6 in.	33 ft. 0 in.	36 ft. 0 in.
Weight of engine in working order		22,000 lb.	25,000 lb.	30,000 lb.	40,000 lb.	45,000 lb.
Capacity of tender tank		500 gals.	600 gals.	800 gals.	900 gals.	1,050 gals.
Weight per yard of lightest steel rail advised		25 lb.	30 lb.	35 lb.	40 lb.	40 lb.
Hauling capacity on a level, in tons of 2,000 lb.		550 tons.	650 tons.	800 tons.	1,000 tons.	1,200 tons.

To compute the hauling capacity on any practicable grade, refer to Table II., page 47.

NOTE.—Refer to page 46 for explanation of hauling capacity; for regular work locomotives should be used at one-half or two-thirds of their full capacity.

NOTE.—The 9½ x 14 cylinders are placed slightly inclined.

LIGHT FOUR-WHEEL-CONNECTED TANK LOCOMOTIVE.

These engines are designed for special service, contractor's work, and other work where the run is not long, on wide or narrow gauge, where a simple design with power is needed without special capacity for speed. The 8 x 14 and 9 x 14 are useful for light work on wide gauge; smaller than 7 x 12 is rarely advisable on wide gauge. The 5 x 10 is adapted for very narrow gauges, and is only advisable for easy work. These engines are well balanced and easy in their motion, being equalized across at front drivers. They are adapted to sharp curves and heavy grades. The proper speed with load is 6 to 10 miles per hour

Cylinders	diameter	5 inches.	6 inches.	7 inches.	8 inches.	9 inches.
	stroke	10 inches.	10 inches.	12 inches.	14 inches.	14 inches.
Diameter of driving wheels		22 inches.	23 inches.	24 inches.	28 inches.	30 inches.
Wheel-base		4 ft. 0 in.	4 ft. 0 in.	4 ft. 8 in.	5 ft. 0 in.	5 ft. 3 in.
Length over all		10 ft. 0 in.	11 ft. 0 in.	12 ft. 7 in.	14 ft. 0 in.	15 ft. 1 in.
Weight in working order (all on drivers)		8,500 lb.	12,000 lb.	15,000 lb.	18,000 lb.	22,000 lb.
Capacity of saddle tank		125 gals.	150 gals.	200 gals.	250 gals.	325 gals.
Weight per yard of lightest steel rail advised		14 lb.	16 lb.	20 lb.	25 lb.	30 lb.
Hauling capacity on a level, in tons of 2,000 lb.		175 tons.	275 tons.	375 tons.	450 tons.	550 tons.

To compute the hauling capacity on any practicable grade, refer to Table I., page 47.

NOTE.—Refer to page 46 for explanation of hauling capacity; for regular work locomotives should be used at one-half or two-thirds of their full capacity.

For actual performances, see WORKING REPORTS on pages 93, 110, to 125, and 132 to 139.

LIGHT FOUR-WHEEL-CONNECTED LOCOMOTIVE, WITH TENDER.

These engines are designed for the same general service as the engines shown on the opposite page, but where the rail requires a lighter engine, or the length of the run renders a tender desirable. A four-wheel tender is sufficient.

A four-wheel locomotive must necessarily have an uneven motion, for a perfect equalizing of the weight is impossible. Except for slow speed and short runs, the locomotives shown on pages 20, 37 and 36 are preferable to the four-wheel-connected style.

		6 inches.	7 inches.	8 inches.	9 inches.
Cylinders	diameter	6 inches.	7 inches.	8 inches.	9 inches.
	stroke	10 inches.	12 inches.	14 inches.	14 inches.
Diameter of driving wheels		26 inches.	28 inches.	30 inches.	33 inches.
Wheel-base of engine		4 ft. 0 in.	4 ft. 8 in.	5 ft. 0 in.	5 ft. 3 in.
Wheel-base of engine and tender		13 ft. 9 in.	14 ft. 6 in.	17 ft. 3 in.	17 ft. 6 in.
Length over all of engine and tender		20 ft. 0 in.	21 ft. 6 in.	24 ft. 6 in.	25 ft. 6 in.
Weight of engine in working order		10,000 lb.	13,000 lb.	16,000 lb.	19,000 lb.
Capacity of tender tank		250 gals.	300 gals.	300 gals.	500 gals.
Weight per yard of lightest steel rail advised		12 lb.	16 lb.	20 lb.	25 lb.
Hauling capacity on a level, in tons of 2,000 lb.		225 tons.	325 tons.	400 tons.	500 tons.

To compute the hauling capacity on any practicable grade, refer to Table II., page 47.

NOTE.—Refer to page 46 for explanation of hauling capacity; for regular work locomotives should be used at one-half or two-thirds of their full capacity.

For actual performances, see WORKING REPORTS on pages 117, 118, 137 and 139.

SIX-WHEEL-CONNECTED MINE LOCOMOTIVE.

		8 inches.	9 inches.	9½ inches
Cylinders { diameter		8 inches.	9 inches.	9½ inches
Cylinders { stroke		14 inches.	14 inches.	14 inches.
Diameter of driving wheels		24 inches.	26 inches.	26 inches.
Wheel-base		5 ft. 5 in.	5 ft. 10 in.	7 ft. 3 in.
Length over all		15 ft. 0 in.	16 ft. 0 in.	17 ft. 0 in.
Extreme width on 36 inches gauge		65 inches.	67 inches.	67 inches.
Extreme height from rail, least advised		6 ft. 6 in.	6 ft. 6 in.	6 ft. 6 in.
Extreme height from rail, least possible		5 ft. 6 in.	5 ft. 8 in.	5 ft. 10 in.
Weight in working order		19,000 lb.	23,000 lb.	26,000 lb.
Capacity of saddle tank		250 gals.	300 gals.	350 gals.
Weight per yard of lightest steel rail advised		16 lb.	20 lb.	25 lb.

Hauling capacity on a level, in tons of

2,000 lb. .... 475 tons.   600 tons.   675 tons.

To compute the hauling capacity on any practicable grade, refer to Table I., page 47.

For practicable hints as to operating mine locomotives, see page 71.

These engines are equalized between the rear and centre drivers and across at the front drivers. The centre drivers are without flanges. It is possible for these engines to pass around curves of 30 to 50 feet radius; but we advise 75 feet as the shortest radius desirable. For the best draft of stack and convenience of engineer, all the height possible should be given. By altering patterns, we can build lower than the lesser height given on opposite page. The widest point is at the cylinders, about two feet above the track. The cab and tank are generally rounded at the top, but may be made to suit the shape of the mine opening.

The weights of these locomotives may be modified, and different sizes of driving wheels used to suit special requirements.

We are prepared to build larger or smaller locomotives of this style, and also for any practicable gauge of track.

NOTE.—Refer to page 46 for explanation of hauling capacity; for regular work locomotives should be used at one-half or two-thirds of their full capacity, and the lesser proportion is advised when the cars have loose wheels.

For actual performances, see WORKING REPORTS on pages 129 and 130.

FOUR-WHEEL-CONNECTED MINE LOCOMOTIVE.

Cylinders { diameter ..	5 inches.	6 inches.	7 inches	8 inches.	9 inches.	10 inches.
{ stroke	10 inches.	10 inches.	12 inches.	14 inches.	14 inches.	14 inches.
Diameter of driving wheels..............	22 inches	23 inches.	24 inches.	26 inches.	28 inches.	28 inches.
Wheel-base	4 ft. 0 in.	4 ft. 0 in.	4 ft. 8 in.	5 ft. 0 in.	5 ft. 3 in.	4 ft. 6 in.
Length over all.......	10 ft. 0 in.	11 ft. 6 in.	12 ft. 7 in.	13 ft. 0 in.	15 ft. 1 in.	16 ft. 9 in.
Extreme width on 36 inches gauge........	60 inches.	62½ in.	64 inches.	65½ in.	67 inches.	68 inches.
Extreme height from rail, least advised ...	5 ft. 0 in.	5 ft. 3 in.	5 ft. 6 in.	6 ft. 4 in.	6 ft. 6 in.	6 ft. 9 in.
Extreme height from rail, least possible...	4 ft. 6 in.	4 ft. 9 in.	5 ft. 0 in.	5 ft. 8 in.	5 ft. 10 in.	5 ft. 10 in.
Weight in working order.................	8,000 lb.	11,500 lb.	15,000 lb.	18,000 lb.	21,500 lb.	24,500 lb.
Capacity of saddle tank	125 gals.	150 gals.	200 gals.	250 gals.	325 gals.	400 gals.
Weight per yard of lightest steel rail advised	16 lb.	16 lb.	20 lb.	25 lb.	30 lb.	30 lb.
Hauling capacity on a level, in tons of 2,000 lb.............. .	150 tons.	250 tons.	350 tons.	425 tons.	525 tons.	600 tons.

To compute the hauling capacity on any practicable grade, refer to Table I., page 47.

NOTE.—Refer to page 46 for explanation of hauling capacity; for regular work locomotives should be used at one-half or two-thirds of their full capacity, and the lesser proportion is advised when the cars have loose wheels.

For actual performances, see WORKING REPORTS, on pages 126 to 131.

For practical hints as to operating mine locomotives, see page 71.

The locomotives illustrated on the opposite page have a cross equalizer at the front drivers. They can pass around curves of 30 feet, or even less radius; but we advise 50 feet as the very shortest radius desirable. By altering patterns we can build lower than the lesser height given on opposite page. For the best draft of stack and convenience of engineer, all the height possible should be given. The widest point is at the cylinders, about two feet above the track. The tank and cab may be of shape required by mine opening.

The weights of these locomotives may be modified, and different sizes of driving wheels used to suit special requirements.

We are prepared to build four-wheel-connected mine locomotives with the same size of cylinders and weights as the locomotives on page 24. We are also prepared to build for any practicable gauge of track.

INSIDE CONNECTED MINE LOCOMOTIVE.

The illustration given below shows our mine locomotive with crank axle and inside cylinders. This construction is expensive and objectionable, and only recommended for very narrow tunnels which cannot be widened at reasonable expense.

We build but one size of this design, with dimensions as follows:

> Cylinders 9 inches diameter by 12 inches stroke.
> Wheel-base 4 feet.
> Length over all 15 feet 6 inches.
> Extreme width on 36 inches gauge 50 inches.
> Least height 5 feet.
> Saddle tank 250 gals.
> Weight in running order 16,000 lbs.
> Lightest rail 25 lbs.
> Hauling capacity on a level 375 tons.

A narrower gauge than 36 inches requires change in patterns for this design.

FOUR-WHEEL-CONNECTED REAR TANK MOTOR.

This design may be built with pilots or with dash-boards, or without either ; and with or without side-flaps, as preferred.

For a more complete description of construction and details, and for practical hints for operating our motors, see pages 61 to 66.

For additional designs of enclosed motors see pages 42, 43, 44 and 45.

MEMORANDUM.—The mine engine 32 of previous editions is now found on page 31.

(WITH PILOTS AND SIDE-FLAPS.)

Cylinders { diameter	6 inches.	7 inches.	8 inches.	9 inches.	10 inches.
{ stroke	10 inches.	12 inches.	14 inches.	14 inches.	14 inches.
Diameter of driving wheels.. ...	23 inches.	24 inches.	28 inches.	30 inches.	33 inches.
Wheel-base	4 ft. 0 in.	4 ft. 8 in.	5 ft. 0 in.	5 ft. 3 in.	4 ft. 6 in.
Length over all	15 ft. 0 in.	15 ft. 6 in.	16 ft. 6 in.	17 ft. 6 in.	18 ft. 6 in.
Height over all	9 ft. 5 in.	9 ft. 5 in.	9 ft. 9 in.	10 ft. 0 in.	10 ft. 2 in.
Weight in working order, all on drivers	14,000 lb.	17,000 lb.	20,000 lb.	25,000 lb.	28,000 lb.
Capacity of rear tank	125 gals.	150 gals.	200 gals.	300 gals.	350 gals.
Weight per yard of lightest steel T rail advised	16 lb.	20 lb.	25 lb.	30 lb.	35 lb.
Hauling capacity on a level, in tons of 2,000 lb.	250 tons.	350 tons.	450 tons.	550 tons.	675 tons.

To compute the hauling capacity on any practicable grade, refer to Table I., page 47.

NOTE.—Refer to page 46 for explanation of hauling capacity; for regular work motors should be used at one-half or two-thirds of their full capacity, and the lesser proportion is advised.

For actual performances, see WORKING REPORTS, pages 92 to 95.

FOUR-WHEEL-CONNECTED SADDLE TANK MOTOR.

This design may be built with pilots or with dash-boards, or without either; and with or without side-flaps, as preferred.

For a more complete description of construction and details, and for practical hints for operating our motors, see pages 61 to 66.

For additional designs of enclosed motors see pages 42, 43, 44 and 45.

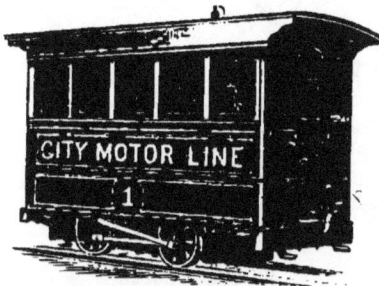

| (WITHOUT DASH-BOARDS OR PILOTS OR SIDE-FLAPS.) | (WITH DASH-BOARDS AND SIDE-FLAPS) |

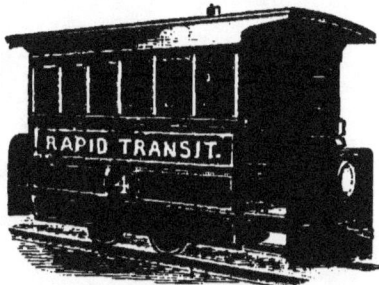

Cylinders { diameter...	6 inches.	7 inches.	8 inches.	9 inches.	10 inches.	12 inches.
Cylinders { stroke.....	10 inches.	12 inches.	14 inches.	14 inches.	14 inches.	18 inches.
Diameter of driving wheels..............	23 inches.	24 inches.	28 inches.	30 inches.	33 inches.	36 inches.
Wheel-base............	4 ft. 0 in.	4 ft. 8 in.	5 ft. 0 in.	5 ft. 3 in.	4 ft. 6 in.	5 ft. 9 in.
Length over all, without pilots or dashboards.	12 ft. 0 in.	13 ft. 0 in.	14 ft. 0 in.	15 ft. 6 in.	18 ft. 0 in.	20 ft. 0 in.
Height over all.......	9 ft. 5 in.	9 ft. 5 in.	9 ft. 9 in.	10 ft. 0 in.	10 ft. 2 in.	11 ft. 0 in.
Weight in working order (all on drivers)..	14,000 lb.	17,000 lb.	20,000 lb.	25,000 lb.	28,000 lb.	40,000 lb.
Cap'ty of saddle tank.	125 gals.	200 gals.	250 gals.	325 gals.	400 gals.	750 gals.
Weight per yard of lightest steel T rail advised	16 lb.	20 lb.	25 lb.	30 lb.	35 lb.	40 lb.
Hauling capacity on a level, in tons of 2,000 lb.................	250 tons.	350 tons.	450 tons.	550 tons.	675 tons.	950 tons.

To compute the hauling capacity on any practicable grade, refer to Table I., page 47.

NOTE.—Refer to page 46 for explanation of hauling capacity; for regular work motors should be used at one-half or two-thirds of their full capacity, and the lesser proportion is advised.

For actual performances, see WORKING REPORTS, pages 92 to 95.

FOUR-WHEEL-CONNECTED PLANTATION LOCOMOTIVE.

These locomotives may be built with wooden cabs (like that of the locomotive on page 26), but for plantation railroads the open sheet-iron canopy is preferable. The water is carried in two rear tanks, one at each side with connecting pipe, and serving as seats for the engineer. This construction gives the lightest possible weight, making this design suitable for light rails or light trestle-work. One pair of driving wheels is equalized across and the locomotive is as free from rocking or oscillating motion as is possible for a four-wheel locomotive. This style is the simplest and least expensive, and is advisable unless the length of the road demands speed or greater fuel and water room. Page 38 shows this general style with tank on boiler.

Cylinders	diameter	5 inches.	6 inches.	7 inches.	8 inches.	9 inches.
	stroke	10 inches.	10 inches.	12 inches.	14 inches.	14 inches.
Diameter of driving wheels		22 inches.	22 inches.	24 inches.	26 inches.	28 inches.
Wheel-base		4 ft. 0 in.	4 ft. 0 in.	4 ft. 8 in.	5 ft. 0 in.	5 ft. 3 in.
Length over all		10 ft. 0 in.	11 ft. 0 in.	12 ft. 7 in.	13 ft. 0 in.	15 ft. 1 in.
Weight in working order (all on drivers)		8,000 lb.	10,500 lb.	14,000 lb.	17,000 lb.	21,000 lb.
Capacity of tank		100 gals.	125 gals.	150 gals.	200 gals.	250 gals.
Weight per yard of lightest steel rail advised		12 lb.	14 lb.	16 lb.	20 lb.	25 lb.
Hauling capacity on a level, in tons of 2,000 lb.		150 tons.	250 tons.	350 tons.	450 tons.	550 tons.

To compute the hauling capacity on any practicable grade, refer to Table I., page 47.

NOTE.—Refer to page 46 for explanation of hauling capacity; for regular work locomotives should be used at one-half or two-thirds of their full capacity.

For actual performances, see WORKING REPORTS on pages 111, 112 and 132.

PLANTATION LOCOMOTIVE, WITH BACK TRUCK.

These locomotives may be built with wooden cabs (as shown on page 37), but for plantation railroads the open sheet-iron canopy is usually preferable. The position of the tank at the rear instead of over the boiler involves some loss of power, but distributes the weight so as to admit the use of a lighter rail. The driving wheels are equalized and a very easy motion secured. Larger driving wheels may be used if greater speed is desired. Very sharp curves are admissible. The fuel is carried in the space over the tank. For very long roads with limited water supply an additional tank on the boiler may be used.

Cylinders { diameter	6 inches.	7 inches.	8 inches.	9 inches.	9½ inches
Cylinders { stroke	10 inches.	12 inches.	14 inches.	14 inches.	14 inches.
Diameter of driving wheels	24 inches.	28 inches.	30 inches.	33 inches.	36 inches.
Diameter of truck wheels	14 inches.	16 inches.	18 inches.	20 inches.	22 inches.
Rigid wheel-base	4 ft. 0 in.	4 ft. 8 in.	5 ft. 0 in.	4 ft. 6 in.	4 ft. 6 in.
Total wheel-base	9 ft. 0 in.	10 ft. 0 in.	10 ft. 6 in.	13 ft. 4 in.	13 ft.10 in.
Length over all	14 ft. 6 in.	17 ft. 0 in.	17 ft. 9 in.	21 ft. 0 in.	22 ft. 0 in.
Weight in working order	14,500 lb.	18,000 lb.	21,500 lb.	26,000 lb.	29,000 lb.
Weight on driving wheels	9,000 lb.	12,000 lb.	14,500 lb.	18,000 lb.	20,000 lb.
Weight on two-wheel radial-bar truck	5,500 lb.	6,000 lb.	7 000 lb.	8,000 lb.	9,000 lb.
Capacity of tank	125 gals.	175 gals.	250 gals.	300 gals.	350 gals.
Weight per yard of lightest steel rail advised	12 lb.	16 lb.	20 lb.	25 lb.	25 lb.
Hauling capacity on a level, in tons of 2,000 lb.	200 tons.	300 tons.	375 tons.	450 tons.	525 tons.

To compute the hauling capacity on any practicable grade, refer to Table II., page 47.

NOTE.—Refer to page 46 for explanation of hauling capacity; for regular work locomotives should be used at one-half or two-thirds of their full capacity.

For actual performances, see WORKING REPORTS on pages 111, 116, 118 and 119.

FORNEY LOCOMOTIVE.

This design was invented and patented by Mr. M. N. Forney. It is advisable, instead of the style shown on page 37, for locomotives of such size that the water and fuel cannot be carried on a two-wheel truck. It may often be used in the place of the locomotives on pages 8, 10, 12, and 14, and is essentially the type shown on pages 25 and 27, modified by connecting the engine and tender in one rigid frame. It is a very simple and efficient design, and capable of a wide range of work, being powerful enough for freight and fast enough for passenger work. If run with the truck ahead, it is, so far as ease of motion and speed are concerned, like the familiar eight-wheel passenger engine (page 4). The driving wheels are equalized, and, except for roads with no sharp curves, the truck is fitted with swinging links. It is adapted to all gauges, and this style, and those with the two-wheel truck, are almost the only ones practicable, unless for very small locomotives, for the 24 inches and other extremely narrow gauges.

We are prepared to modify this design by adding a two-wheel front truck, but do not recommend it, as it makes too long an engine with too little power.

Cylinders { diameter	9 inches.	9½ inches	10 inches.	12 inches.	14 inches.
{ stroke	14 inches.	14 inches.	16 inches.	18 inches.	20 inches.
Diameter of driving wheels	33 to 36 in.	36 to 40 in.	40 to 44 in.	44 to 48 in.	48 inches.
Diameter of truck wheels	18 to 20 in.	20 to 22 in.	22 inches.	24 inches.	24 inches.
Rigid wheel-base	4 ft. 6 in.	4 ft. 6 in.	5 ft. 3 in.	5 ft. 9 in.	7 ft. 0 in.
Total wheel-base	16 ft. 3 in.	17 ft. 0 in.	17 ft. 10 in.	18 ft. 2 in.	19 ft. 6 in.
Length over all, including pilot	28 ft. 0 in.	29 ft. 6 in.	30 ft. 6 in.	32 ft. 0 in.	34 ft. 0 in.
Weight in working order	30,000 lb.	33,000 lb.	38,000 lb.	48,000 lb.	58,000 lb.
Weight on driving wheels	18,000 lb.	20,500 lb.	25,000 lb.	33,000 lb.	40,000 lb.
Weight on four-wheel rear truck	12,000 lb.	12,500 lb.	13,000 lb.	15,000 lb.	18,000 lb.
Water capacity of tank	400 gals.	450 gals.	500 gals.	700 gals.	900 gals.
Weight per yard of lightest steel rail advised	25 lb.	25 lb.	30 lb.	35 lb.	45 lb.

Hauling capacity on a level, in tons of 2,000 lb.	450 tons.	525 tons.	675 tons.	850 tons.	1,050 tons.

To compute the hauling capacity on any practicable grade, refer to Table II , page 47.

NOTE.—Refer to page 46 for explanation of hauling capacity; for regular work locomotives should be used at one-half or two-thirds of their full capacity.

NOTE.—The 9 x 14 and 9½ x 14 cylinders are placed slightly inclined.

For actual performances, see WORKING REPORTS on pages 87, 89, 96, 97 and 143.

BACK TRUCK PLANTATION LOCOMOTIVE, WITH WOODEN CAB.

This design is the same as the Back Truck Plantation Locomotive described on page 35, with the exception of a wooden cab. For these small sizes the two-wheel radial truck is preferable to the four wheel, as it admits ample fuel and water capacity, and is simpler and can pass sharper curves. For very long roads, or where the water supply is limited, an additional tank may be carried on the boiler, but this is advisable only in exceptional cases. These locomotives are desirable for plantation roads, or other roads with light or portable track, where the open canopy is not preferable, and where the saddle-tank style is not desired. For light passenger service, if extra speed is needed, larger driving wheels may be used. The weight is well distributed, the motion very easy, and sharp curves admissible.

Cylinders { diameter	6 inches.	7 inches.	8 inches.	9 inches.	9½ inches
{ stroke	10 inches.	12 inches.	14 inches.	14 inches.	14 inches.
Diameter of driving wheels	24 inches.	28 inches.	30 inches.	33 inches.	36 inches.
Diameter of truck wheels	14 inches.	16 inches.	18 inches.	20 inches.	22 inches.
Rigid wheel-base	4 ft. 0 in.	4 ft. 8 in.	5 ft. 0 in.	4 ft. 6 in.	4 ft. 6 in.
Total wheel-base	9 ft. 0 in.	10 ft. 0 in.	10 ft. 6 in.	13 ft. 4 in.	13 ft. 10 in.
Length over all	16 ft. 6 in.	19 ft. 0 in.	19 ft. 9 in.	20 ft. 0 in.	21 ft. 0 in.
Weight in working order	15,000 lb.	18,500 lb.	21,500 lb.	26,000 lb.	29,000 lb.
Weight on driving wheels	9,000 lb.	12,000 lb.	14,500 lb.	18,000 lb.	20,000 lb.
Weight on two-wheel radial-bar truck	6,000 lb.	6,500 lb.	7,000 lb.	8,000 lb.	9,000 lb.
Capacity of tank	125 gals.	175 gals.	250 gals.	300 gals.	350 gals.
Weight per yard of lightest steel rail advisable	16 lb.	16 lb.	20 lb.	25 lb.	25 lb.
Hauling capacity on a level, in tons of 2,000 lb.	200 tons.	300 tons.	375 tons.	450 tons.	525 tons.

To compute the hauling capacity on any practicable grade, refer to Table II., page 47.

NOTE.—Refer to page 46 for explanation of hauling capacity; for regular work locomotives should be used at one-half or two-thirds of their full capacity.

For actual performances, see WORKING REPORTS on pages 86, 115, 119 and 136.

LIGHT FOUR-WHEEL-CONNECTED TANK LOCOMOTIVE, WITH OPEN CANOPY.

This design is identical with that on page 26, with the exception of the open sheet-iron canopy, and is also identical with that on page 34, with the exception of the position of the water tank. The open canopy is cheaper than a wooden cab, and generally preferable for hot climates; the saddle tank, except for very light rails, is preferable to rear tank, as it has more capacity and increases the total weight. For the three smallest sizes solid chilled iron wheels may be used, and are cheaper than steel tires. These locomotives are well balanced and the greatest ease of motion possible for a four wheel locomotive is secured by a cross equalizer at the front springs. They are adapted to sharp curves and steep grades. The proper speed with load is 6 to 10 miles per hour. Smaller than 7 x 12 cylinders of this style is rarely advisable for wide gauge. This style may also be built with separate tender, like page 27.

Cylinders { diameter	5 inches.	6 inches.	7 inches.	8 inches.	9 inches.
{ stroke	10 inches.	10 inches.	12 inches.	14 inches.	14 inches.
Diameter of driving wheels	23 inches.	23 inches.	24 inches.	28 inches.	30 inches.
Wheel-base	4 ft. 0 in.	4 ft. 0 in.	4 ft. 8 in.	5 ft. 0 in.	5 ft. 3 in.
Length over all	10 ft. 0 in.	11 ft. 0 in.	12 ft. 7 in.	14 ft. 0 in.	15 ft. 1 in.
Weight in working order (all on drivers)	8,500 lb.	12,000 lb.	15,000 lb.	18,000 lb.	22,000 lb.
Capacity of saddle tank	125 gals.	150 gals.	200 gals.	250 gals.	325 gals.
Weight per yard of lightest steel rail advised	12 lb.	16 lb.	16 to 20 lb.	25 lb.	30 lb.
Hauling capacity on a level, in tons of 2,000 lb.	175 tons.	275 tons.	350 tons.	450 tons.	550 tons.

To compute the hauling capacity on any practicable grade, refer to Table I., page 47.

NOTE.—Refer to page 46 for explanation of hauling capacity; for regular work locomotives should be used at one-half or two-thirds of their full capacity.

For actual performances, see WORKING REPORTS on pages 110 to 125, and 133.

LIGHT BACK TRUCK LOCOMOTIVE WITH OPEN CANOPY.

(FOR LOGGING AND PLANTATION ROADS AND SIMILAR SERVICE.)

This design is the same as that on pages 20 and 21, with the exception of the open sheet iron canopy, which is cheaper than the wooden cab and better adapted for hot climates. We are prepared to build larger sizes with canopy. For the smaller sizes solid chilled wheels may be used instead of steel tire, and are cheaper. The driving wheels are equalized and the motion easy, even on rough track, while the large proportion of weight on the driving wheels secures power. The truck is centre bearing, with swing motion and radial bar. Sharp curves, light rails, steep grades, and heavy loads, and, when needful, fast speeds are practicable. These locomotives are specially adapted to logging roads, plantation roads, and other service where there are objections to the four-wheel locomotive.

Cylinders	diameter	6 inches.	7 inches.	8 inches.	9 inches.	9½ inches
	stroke	10 inches.	12 inches.	14 inches.	14 inches.	14 inches.
Diameter of driving wheels		24 inches.	28 inches.	30 inches.	33 inches.	36 inches.
Diameter of truck wheels		16 inches.	16 inches.	18 inches.	20 inches.	2 inches.
Rigid wheel-base		4 ft. 0 in.	4 ft. 8 in.	5 ft. 0 in.	4 ft. 6 in.	4 ft. 6 in.
Total wheel-base		8 ft. 6 in.	9 ft. 1 in.	9 ft. 10 in.	12 ft. 4 in.	12 ft. 6 in.
Length over all		14 ft. 0 in.	16 ft. 4 in.	16 ft. 9 in.	20 ft. 0 in.	21 ft. 0 in.
Weight in working order		14,000 lb.	18,000 lb.	22,000 lb.	27,000 lb.	30,000 lb.
Weight on driving wheels		10,500 lb.	13,500 lb.	17,000 lb.	20,000 lb.	23,000 lb.
Weight on two-wheel radial-bar truck		3,500 lb.	4,500 lb.	5,000 lb.	7,000 lb.	7,000 lb.
Capacity of saddle-tank		150 gals.	200 gals.	250 gals.	375 gals.	400 gals.
Weight per yard of lightest steel rail advised		16 lb.	16 lb.	20 lb.	25 lb.	25 lb.
Hauling capacity on a level, in tons of 2,000 lb.		225 tons.	325 tons.	425 tons.	500 tons.	575 tons.

To compute the hauling capacity on any practicable grade, refer to Table I., page 47.

NOTE.—Refer to page 46 for explanation of hauling capacity; for regular work locomotives should be used at one-half or two-thirds of their full capacity.

NOTE.—The proportions of the 9 x 14 and 9½ x 14 locomotives differ slightly from the illustration above.

For actual performances, see WORKING REPORTS on pages 133, 135, 136 and 137.

MILL LOCOMOTIVES.

This design is the same as that described on pages 34, 38 and 41, modified for use inside of mills. These locomotives are used for moving hot ingots and blooms to the rolls, and no cab is required when the locomotive runs wholly inside the mill. When the locomotive is used about a Bessemer converter, for hauling fluid metal or taking ingots from the pit, or for moving cinder from the blast furnace, a sheet iron cab is desirable as shown on the opposite page. The 5 x 10 and 6 x 10 locomotives are used at the rolls, and the larger sizes are generally advisable for cinder and ingot work. For the larger sizes steel tired wheels are desirable, but for the smaller sizes solid chilled wheels may be preferable. Usually no bell is needed.

MEMORANDUM.—The water may be carried in a saddle tank like the cut "Bloom," or in two-connected rear tanks like the "Ingot." The weight and power of the saddle tank design is slightly the greater. The rear tank design gives a slightly better outlook for the engineer.

Cylinders { diameter	5 inches.	6 inches.	7 inches.	8 inches.	9 inches.
{ stroke	10 inches.	10 inches.	12 inches.	14 inches.	14 inches.
Diameter of driving wheels	22 inches.	22 inches.	22 inches.	24 inches.	28 inches.
Wheel-base	4 ft. 0 in.	4 ft. 0 in.	4 ft. 8 in.	5 ft. 0 in.	5 ft. 3 in.
Length over all	10 ft. 0 in.	11 ft. 0 in.	12 ft. 7 in.	13 ft. 0 in.	15 ft 1 in.
Weight in working order, with two rear tanks (all on drivers)	7,500 lb.	10,000 lb.	14,000 lb.	17,000 lb.	21,000 lb.
Capacity of two tanks placed at rear	100 gals.	125 gals.	150 gals.	200 gals.	250 gals.
Hauling capacity on a level, in tons of 2,000 lb.	150 tons.	250 tons.	350 tons.	450 tons.	550 tons.

To compute the hauling capacity on any practicable grade, refer to Table I., page 47.

NOTE.—Refer to page 46 for explanation of hauling capacity; for regular work locomotives should be used at one-half or two-thirds of their full capacity.

For actual performances, see WORKING REPORTS on pages 110 to 125.

STEEL WORKS AND COKE OVEN LOCOMOTIVES.

This design is like pages 26 and 38, with the details arranged to suit the special requirements. Very little wood-work is used, and the cab is made of sheet steel and shaped so as to clear any obstructions, and also to protect the engineer from heat. For Bessemer converters the cab is usually closed except at one side; for cinder and ingot work the cab may be opened at the sides and closed at the front and back; for miscellaneous work about mills and furnaces an open canopy like pages 38 and 34 may be preferable; for coke ovens, where the locomotives haul the larries on a track placed between two rows of ovens, the cab is usually closed at the front, partly closed at the sides and open at the back.

Cylinders { diameter	5 inches.	6 inches.	7 inches.	8 inches.	9 inches.
{ stroke	10 inches.	10 inches.	12 inches.	14 inches.	14 inches.
Diameter of driving wheels	22 inches.	22 inches.	24 inches.	28 inches.	30 inches.
Wheel-base	4 ft. 0 in.	4 ft. 0 in.	4 ft. 8 in.	5 ft. 0 in.	5 ft. 3 in.
Length over all	10 ft. 0 in.	11 ft. 0 in.	12 ft. 7 in.	13 ft. 0 in.	15 ft. 1 in.
Weight in working order (all on drivers)	8,500 lb.	12,000 lb.	15,000 lb.	18,000 lb.	22,000 lb.
Capacity of saddle tank	125 gals.	150 gals.	200 gals.	250 gals.	325 gals.
Hauling capacity on a level, in tons of 2,000 lb.	175 tons.	275 tons.	350 tons.	450 tons.	550 tons.

To compute the hauling capacity on any practicable grade, refer to Table I., page 47.

NOTE.—Refer to page 46 for explanation of hauling capacity; for regular work locomotives should be used at one-half or two-thirds of their full capacity.

For actual performances, see WORKING REPORTS on pages 110 to 125.

LIGHT BACK TRUCK MOTOR.

(WITH SADDLE TANK.)

This design may be built with pilots or with dash-boards or without either ; and with or without side-flaps, as preferred.

For a more complete description of construction and details, and for practical hints for operating our motors, see pages 61 to 66.

(WITH DASH-BOARDS AND SIDE-FLAPS.)

Cylinders { diameter	7 inches.	8 inches.	9 inches.	10 inches.
{ stroke	12 inches.	14 inches.	14 inches.	14 inches.
Diameter of driving wheels	28 inches.	30 inches.	33 inches.	33 inches.
Diameter of truck wheels	16 inches.	18 inches.	20 inches.	20 inches
Rigid wheel-base	4 ft. 8 in.	5 ft. 0 in.	4 ft. 6 in	4 ft. 6 in.
Total wheel-base	8 ft. 3 in.	8 ft. 9 in.	9 ft. 3 in.	9 ft. 3 in.
Length over all	15 ft. 6 in.	16 ft. 0 in.	17 ft. 6 in.	17 ft 6 in.
Height over all	9 ft. 5 in.	9 ft. 9 in.	10 ft. 0 in.	10 ft. 0 in.
Total weight in working order	19,000 lb.	23,000 lb.	28,000 lb.	31,500 lb.
Weight on driving wheels	14,000 lb.	17,000 lb.	21,500 lb.	24,000 lb.
Weight on two-wheel radial-bar truck	5,000 lb.	6 000 lb.	6,500 lb.	7,500 lb.
Capacity of saddle tank	200 gals.	250 gals.	325 gals	400 gals.
Weight per yard of lightest steel T rail advised	16 to 20 lb.	25 lb.	30 lb.	30 lb.
Hauling capacity on a level, in tons of 2,000 lb...	350 tons.	425 tons.	525 tons.	625 tons.

To compute the hauling capacity on any practicable grade, refer to Table I., page 47.

NOTE.—Refer to page 46 for explanation of hauling capacity; for regular work motors should be used at one-half to two-thirds of their full capacity, and the lesser proportion is advised.

For actual performances, see WORKING REPORTS on pages 92 to 94.

BACK-TRUCK MOTOR.

(WITH SADDLE TANK.)

This design may be built with pilots, or with dash-boards, or without either ; and with or without side-flaps, as preferred.

For a more complete description of construction and details, and for practical hints for operating our motors, see pages 61 to 66.

(WITH PILOTS, WITHOUT SIDE-FLAPS.)

Cylinders { diameter	10 inches.	12 inches.	14 inches.
stroke	16 inches.	18 inches.	20 inches.
Diameter of driving wheels.	36 inches.	40 inches.	44 inches.
Diameter of truck wheels.................	22 inches.	24 inches	26 inches.
Rigid wheel-base............................	5 ft. 3 in.	5 ft. 9 in.	6 ft. 3 in.
Total wheel-base............................	11 ft. 3 in.	11 ft. 9 in.	13 ft. 0 in.
Length over all.................	19 ft. 0 in.	19 ft. 6 in.	21 ft. 0 in.
Height over all............................ ...	10 ft. 3 in.	11 ft. 0 in.	11 ft. 3 in.
Total weight in working order.......	35,000 lb.	43,000 lb.	54,000 lb.
Weight on driving wheels........	28,000 lb.	35,000 lb.	44,000 lb.
Weight on two-wheel radial-bar truck..........	7,000 lb.	8,000 lb.	10,000 lb.
Capacity of saddle tank.	500 gals.	750 gals.	900 gals.
Weight per yard of lightest steel T rail advised.	30 lb.	35 lb.	40 lb.
Hauling capacity on a level, in tons of 2,000 lb................	700 tons.	900 tons.	1,100 tons.

To compute the hauling capacity on any practicable grade, refer to Table I., page 47.

NOTE.—Refer to page 46 for explanation of hauling capacity; for regular work motors should be used at one-half or two-thirds of their full capacity, and the lesser proportion is advised.

For actual performances, see WORKING REPORTS on page 95.

BACK TRUCK MOTOR.

(WITH REAR TANK.)

This design may be built with pilots, or with dash-boards or without either ; and with or without side-flaps, as preferred.

For a more complete description of construction and details, and for practical hints for operating our motors, see pages 61 to 66.

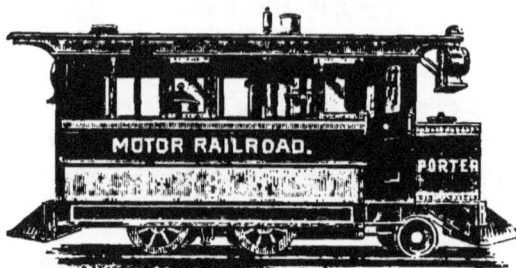

(WITH PILOTS AND SIDE-FLAPS.)

Cylinders diameter	7 inches	8 inches	9 inches	10 inches	10 inches	12 inches	14 inches
Cylinders stroke	12 inches	14 inches	14 inches	11 inches	16 inches	18 inches	20 inches
Diameter of driving wheels	28 inches	30 inches	33 inches	36 inches	36 inches	40 inches	44 inches
Diameter of truck wheels	16 inches	18 inches	20 inches	22 inches	22 inches	24 inches	26 inches
Rigid wheel-base	4 ft. 8 in.	5 ft. 0 in.	5 ft. 3 in.	5 ft. 3 in.	5 ft. 3 in.	5 ft. 9 in.	6 ft. 3 in.
Total wheel-base	8 ft. 5 in.	8 ft. 9 in.	10 ft. 0 in.	10 ft. 0 in.	10 ft. 7 in.	14 ft. 0 in.	15 ft. 0 in.
Length over all (including pilots or dashboards)	18 ft. 0 in.	19 ft. 3 in.	22 ft. 10 in.	23 ft. 11 in	25 ft. 10 in	27 ft. 4 in.	29 ft. 0 in.
Height over all	9 ft. 7 in.	9 ft. 10 in.	10 ft. 0 in.	10 ft. 2 in.	10 ft. 3 in.	11 ft. 2 in.	11 ft. 3 in.
Weight in working order	20,000 lb.	23,000 lb.	28,000 lb.	31,000 lb.	36,000 lb.	44,000 lb.	54,000 lb.
Weight on driving wheels	13,000 lb.	15,000 lb.	19,000 lb.	22,000 lb.	25,000 lb.	32,000 lb.	40,000 lb.
Weight on two-wheel radial-bar truck	7,000 lb.	8,000 lb.	9,000 lb.	9,000 lb.	11,000 lb.	12,000 lb.	14,000 lb.
Water capacity of rear tank	150 gals.	200 gals.	300 gals.	350 gals.	400 gals.	500 gals.	500 gals.
Weight per yard of lightest steel T rail advised	16 lb.	20 to 25 lb	25 lb.	30 lb.	30 lb.	35 lb.	40 lb.
Hauling capacity on a level, in tons of 2,000 lb.	300 tons.	375 tons.	475 tons.	575 tons.	650 tons.	850 tons.	1,000 tons

To compute the hauling capacity on any practicable grade, refer to Table II., page 47.

NOTE.—Refer to page 46 for explanation of hauling capacity; for regular work motors should be used at one-half or two-thirds of their full capacity, and the lesser proportion is advised.

For actual performances, see WORKING REPORTS on pages 94 and 95.

DOUBLE-.ENDER MOTOR.

(WITH PONY TRUCK EACH END, AND SADDLE-TANK.)

This design may be built with pilots, or with dashboards, or without ·either; and with or without side-flaps as preferred. The position of the tank on the boiler is necessary for a proper distribution of weight. This design is especially intended for fast speed.

For a more complete description of construction and details, and for practical hints for operating our motors, see pages 61 to 66.

Cylinders { diameter	8 inches.	9 inches.	10 inches.	12 inches.	14 inches.
Cylinders { stroke	14 inches	14 inches.	16 inches.	18 inches.	20 inches.
Diameter of driving wheels	30 to 33 in.	33 to 36 in.	40 to 44 in.	44 inches.	48 inches.
Diameter of truck wheel	16 to 18 in.	18 to 20 in.	22 to 24 in.	24 inches.	26 inches.
Rigid wheel-base	5 ft. 0 in.	5 ft. 9 in.	6 ft. 6 in.	6 ft. 9 in.	7 ft. 0 in.
Total wheel-base	15 ft. 0 in.	15 ft. 9 in.	18 ft. 6 in.	20 ft. 0 in.	21 ft. 0 in.
Length over all, including pilots.	24 ft. 0 in.	27 ft. 0 in.	32 ft. 0 in.	34 ft. 0 in.	36 ft. 6 in.
Height over all	9 ft. 5 in.	9 ft. 9 in.	10 ft. 0 in.	11 ft. 0 in.	11 ft. 3 in.
Total weight in working order	24,000 lb.	30,000 lb.	40,000 lb.	50,000 lb.	58,000 lb.
Weight on driving wheels	15,000 lb.	19,000 lb.	26,000 lb.	33,000 lb.	40,000 lb.
Weight on two trucks	9,000 lb.	11,000 lb.	14,000 lb.	17,000 lb.	18,000 lb.
Capacity of saddle tank	250 gals.	325 gals.	500 gals.	750 gals.	900 gals.
Weight per yard of lightest steel T rail advised	20 lb.	25 lb.	30 lb.	35 lb.	40 lb.
Hauling capacity on a level, in tons of 2,000 lb.	350 tons.	450 tons.	650 tons.	800 tons.	1,000 tons

To compute the hauling capacity on any practicable grade, refer to Table I., page 47.

NOTE.—Refer to page 46 for explanation of hauling capacity; for regular work motors should be used at one-half or two-thirds of their full capacity, and the lesser proportion is advised.

For actual performances, see WORKING REPORTS on page 94.

HAULING CAPACITY EXPLAINED.

GUARANTEED CAPACITY.

CONDITIONS.

The number of tons given as the hauling capacity of each locomotive is *not* the amount of *freight* it can haul, but is the *total weight* of the heaviest train, including the weight of the cars and of their loads, which we will guarantee the locomotive to start and haul in addition to the locomotive itself (and its tender), on straight track in good condition ; the cars are to be in good order, and of such construction as not to cause unusual friction. The rate of speed is not supposed to be excessive, but only such as the locomotive can attain while doing its best work ; this may be from 5 to 15 miles per hour, according to the design of the locomotive. Over-loaded or empty cars are harder to haul than a train of the same total weight made up of properly loaded cars ; mine cars, especially those with loose wheels, are also hard to haul.

LEVELS.

The *level* mentioned is supposed to be absolute, which is almost unattainable in practice, and we therefore advise that in selecting an engine for work on a so-called " level " road, the capacity of the engine on a 5 or 10 feet per mile grade be taken.

DAILY WORK.

The *regular* work of a locomotive should not exceed one-half to two-thirds of its full capacity. This allowance is advisable for the best economy of operation to provide a surplus of power for special occasions, and to cover the imperfections of track and rolling-stock as found in average practice.

FAVORABLE CONDITIONS.

On *short grades*, where it is not necessary to start the train on the grade, a locomotive can be regularly used to good economy at its full capacity, and often at considerably over its estimated capacity. In such cases the train is taken up the grade by its momentum and the locomotive only helps to keep it in motion. Grades one-quarter of a mile long, when favorably situated, may be thus overcome.

SPEED.

For passenger service the resistance of speed becomes an important element, but no exact rule can be given that will apply to all cases. For very high speeds it may be best not to haul trains of more than one-third or one-half of the full capacity at slow speeds.

TABLES FOR COMPUTING HAULING CAPACITY ON GRADES.

to be used in connection with the hauling capacity on a level given for each locomotive on pages 4 to 45.

In these tables 100 per cent. stands for the hauling capacity on a level. Opposite each grade is given the proper percentage to denote the hauling capacity on that grade. (For fuller explanation, see examples on following pages.)

	TABLE I.			TABLE II.	
	For Saddle Tank Locomotives.			For Locomotives with Tender.	
GRADES.		PERCENTAGES.	GRADES.		PERCENTAGES.
On a level the hauling capacity is		100 per cent.	On a level the hauling capacity is		100 per cent.
1 foot per mile		94 "	1 foot per mile		94 "
2 feet "		90 "	2 feet "		90 "
3 " "		86 "	3 " "		86 "
5 " "		78 "	5 " "		78 "
8 " "		69 "	8 " "		69 "
10 " "		64 "	10 " "		64 "
15 " "		54 "	15 " "		54 "
20 " "		47 "	20 " "		46 "
25 " "		42 "	25 " "		41 "
26$\frac{4}{10}$ "		40 "	26$\frac{4}{10}$ "		39 "
30 " "		37 "	30 " "		36 "
35 " "		33 "	35 " "		32 "
40 " "		30 "	40 " "		29 "
45 " "		28 "	45 " "		27 "
50 " "		26 "	50 " "		25 "
52$\frac{5}{10}$ "		25 "	52$\frac{5}{10}$ "		24 "
55 " "		24 "	55 " "		23 "
60 " "		22 "	60 " "		21 "
65 " "		21 "	65 " "		20 "
70 " "		20 "	70 " "		18 "
75 " "		19 "	75 " "		17 "
80 " "		18 "	80 " "		16 "
85 " "		17 "	85 " "		15 "
90 " "		16 "	90 " "		14 "
95 " "		15 "	95 " "		13 "
100 " "		14 "	100 " "		12 "
105$\frac{5}{10}$ "		13½ "	105$\frac{5}{10}$ "		11½ "
110 " "		13 "	110 " "		11 "
120 " "		12 "	120 " "		10 "
130 " "		11 "	130 " "		9 "
132 " "		10¾ "	132 " "		8¾ "
140 " "		10 "	140 " "		8 "
150 " "		9½ "	150 " "		7½ "
158$\frac{4}{10}$ "		9¼ "	158$\frac{4}{10}$ "		7¼ "
160 " "		9 "	160 " "		7 "
170 " "		8½ "	170 " "		6½ "
180 " "		8 "	180 " "		6 "
184$\frac{2}{10}$ "		7¾ "	184$\frac{2}{10}$ "		5¾ "
190 " "		7½ "	190 " "		5½ "
200 " "		7 "	200 " "		5 "
211$\frac{1}{10}$ "		6½ "	211$\frac{1}{10}$ "		4½ "
225 " "		6 "	225 " "		4 "
250 " "		5½ "	250 " "		3½ "
264 " "		5¼ "	264 " "		3¼ "
275 " "		5 "	275 " "		3 "
300 " "		4½ "	300 " "		2½ "
316$\frac{5}{10}$ "		4¼ "	316$\frac{5}{10}$ "		2¼ "
325 " "		4 "	325 " "		2 "
350 " "		3½ "	350 " "		1½ "
375 " "		3 "	375 " "		1¼ "
400 " "		2¾ "	400 " "		1 "
450 " "		2¼ "			
500 " "		2 "			

DIRECTIONS FOR USING THE PRECEDING TABLES.

I.—To compute how many tons a locomotive can haul up a grade.

With the description and illustration of each locomotive, pages 4 to 45, is given, in tons of 2,000 lbs., its hauling capacity on a level with a reference to Table I. for saddle-tank locomotives, or to Table II, for locomotives with tender. Referring to the proper table, find the grade, and note the percentage given for it. This percentage of the hauling capacity on a level will be the number of tons which the locomotive can haul up the grade.

EXAMPLE I.—What is the hauling capacity up a grade of 300 feet per mile of the 9 x 14 cylinders locomotive, page 26 ?

Page 26 gives the hauling capacity on a level for this locomotive 550 tons. Table I. gives 4½ as the percentage for a 300 feet grade. Four and one-half per cent. of 550 gives (disregarding fractions) 25 tons as the hauling capacity of this locomotive on a 300 feet grade.

EXAMPLE II.—How much can the 12 x 16 cylinders locomotive, page 16, pull up a grade at 50 feet per mile ?

Page 16 states the hauling capacity on a level at 800 tons. Table II. gives 25 as the percentage for a 50 feet grade, and 25 per cent of 800 is 200 tons, the hauling capacity on a 50 feet grade.

II.—To select a locomotive of suitable power for any required work.

Add 50 or 100 per cent. to the regular work to be done, according to the margin of surplus power desired and for allowance for imperfections of track, cars, etc. (See explanation on page 46.) Refer to Table I. or Table II., as the case might be, for the percentage for the given grade. The regular work to be done, as above increased, will then be this percentage of the locomotive's hauling capacity on a level ; and the capacity on a level is found by multiplying by 100, and dividing by the rate of percentage. The locomotive may then be selected from the catalogue according to the nature of the service and the hauling capacity on a level given for each locomotive.

EXAMPLE.—It is desired to haul a load of 150 tons of cars and lading regularly up a grade of 50 feet per mile. What is the smallest saddle-tank locomotive advisable ?

Adding 50 to 100 per cent. to 150 tons gives 225 to 300 tons. Table I. states 26 as the percentage for a 50 feet grade ; 225 multiplied by 100 and divided by 26 gives 866 tons, or 300 multiplied by 100 and divided by 26 gives 1,154 tons. A locomotive of 866 to 1,154 tons capacity on a level is thus indicated, and the catalogue gives a choice between page 24, 12 x 18 cylinders; page 23, 12 x 18 cylinders; and page 21, 12 x 18 cylinders; and it might also be noted that if the load or grade could be slightly reduced, or if the grade were so situated that it could be to a considerable extent overcome by the impetus of the train, a 10 x 16 cylinders locomotive would be available.

MEMORANDA.—*These tables may also be used,* when the hauling capacity of a locomotive on a given grade is known, to compute its hauling capacity on greater or less grades.

Also when a locomotive's capacity on a given grade or on a level is known, to compute the steepest grade up which it can haul any desired practicable load.

LOCATING GRADES. When an elevation is to be overcome it is often possible to secure the greatest economy of operation by retaining an easy gradient as long as possible and then introducing a steep grade, which may be overcome by the momentum of the train; or the train may be divided on the grade, or an extra locomotive may be used as a pusher.

EXCESSIVE GRADES. On very steep grades, say over 300 feet per mile, a wet or slippery rail, or very hard running cars, or other difficulty, may reduce the load an engine can haul in greater proportion than on less grades. It is possible to haul light loads up 600 feet per mile grade with our locomotives; but, from the above reasons, and also on account of the difficulty of controling the engine and train coming down, about 450 feet is about as steep for long grades as is usually practicable. For very heavy grades, engines should be specially designed.

SEE NEXT PAGE. Attention is also called to the Table on page 50 which will show at a glance without requiring any calculation the power of locomotives of different weights on all practicable grades. This table, although not absolutely exact, is very nearly correct and very convenient.

TABLE SHOWING THE HEAVIEST TRAINS LOCOMOTIVES WITH 6,000 TO 50,000 POUNDS WEIGHT ON THE DRIVING WHEELS CAN HAUL ON GRADES FROM A DEAD LEVEL TO 11 FEET PER 100 (580 12/16 FEET PER MILE).

POUNDS WEIGHT ON ALL DRIVING WHEELS.	6,000 lb.	8,000 lb.	10,000 lb.	12,000 lb.	14,000 lb.	16,000 lb.	18,000 lb.	20,000 lb.	22,000 lb.	24,000 lb.	26,000 lb.	28,000 lb.	30,000 lb.	32,000 lb.	34,000 lb.	36,000 lb.	38,000 lb.	40,000 lb.	42,000 lb.	44,000 lb.	46,000 lb.	48,000 lb.	50,000 lb.
Absolute level	165	220	275	330	385	440	495	550	605	660	715	770	825	880	935	990	1045	1100	1155	1210	1265	1320	1375
¼ per cent = 13 3/16 feet per mile	95	127	159	190	222	254	286	319	350	382	414	446	478	510	542	574	606	638	670	702	734	766	798
½ " = 26 4/16 "	66	89	111	133	156	178	200	223	245	267	290	312	334	357	379	401	424	446	468	491	513	535	557
¾ " = 39 8/16 "	51	68	85	102	119	136	153	170	187	204	221	238	255	272	289	306	323	340	357	374	391	408	425
1 " = 52 8/16 "	40	54	67	81	94	108	121	135	149	162	176	189	203	217	230	244	257	271	284	298	311	325	338
1¼ " = 66 "	33	44	56	67	78	90	101	112	124	135	146	158	169	180	192	203	214	225	237	248	259	271	282
1½ " = 79 2/16 "	29	39	48	58	68	77	87	97	106	116	126	135	145	155	164	174	184	193	203	213	223	232	242
1¾ " = 92 8/16 "	25	33	41	50	58	66	74	83	91	99	107	116	124	132	140	149	157	165	174	182	190	198	207
2 " = 105 8/16 "	22	29	37	44	51	59	66	74	81	88	96	103	111	118	125	133	140	148	155	162	170	177	185
2½ " = 132 "	17	23	29	35	41	47	53	59	64	70	76	82	88	94	100	106	112	118	124	130	135	141	147
3 " = 158 8/16 "	14	19	23	28	33	37	42	47	51	56	61	65	70	75	79	84	89	93	98	103	108	112	117
3½ " = 184 10/16 "	12	16	20	24	28	32	36	40	44	48	52	56	60	64	68	72	76	80	84	88	92	96	100
4 " = 211 2/16 "	10	14	17	21	24	28	31	35	38	42	45	49	52	56	59	63	66	70	73	77	80	84	87
4½ " = 237 10/16 "	9	12	15	18	21	24	27	30	33	36	39	42	45	48	51	54	57	60	63	66	69	72	75
5 " = 264 "	8	11	13	16	18	21	24	26	29	31	34	37	39	42	44	47	49	52	55	57	60	62	65
5½ " = 290 4/16 "	7	9	12	14	17	19	22	24	26	29	31	34	36	38	41	43	45	48	50	53	55	57	60
6 " = 316 8/16 "	6	8	10	12	15	17	19	21	23	25	27	29	31	33	35	37	40	42	44	46	48	50	52
7 " = 369 6/16 "	5	7	8	10	12	13	15	17	18	20	22	23	25	27	28	30	32	34	35	37	39	41	42
8 " = 422 4/16 "	4	6	7	9	10	12	13	15	16	18	19	21	22	24	25	27	28	30	31	33	34	36	38
9 " = 475 2/16 "	3	4	5	6	7	8	9	11	12	13	14	15	16	17	18	19	20	22	23	24	25	26	27
10 " = 528 "	2	3	3	4	5	6	7	7	8	9	10	11	11	12	13	14	15	15	16	17	18	19	20
11 " = 580 12/16 "	1	2	2	3	3	4	4	5	5	6	6	7	7	8	8	9	9	10	10	11	11	12	13

NOTE.—The above table is approximate, and for a variety of reasons will not in all cases agree exactly with the hauling capacity given for various locomotives, or with the Tables given on page 47.

SPECIAL CAUTION.

In using the opposite Table it must be borne in mind that locomotives ought not to be worked regularly at over one-half to three-fourths of their full power according to circumstances; also that for saddle-tank locomotives it is safest to reckon the driving weight with the tank about half full; also tender must be counted as a part of the train, and to be exact in case of engines with trucks the weight on the truck should be deducted (on this basis some designs could not ascend the steepest grades even without any train). The weight of train is given in tons of 2,000 lbs., and includes the weight of *cars and their loads.* The friction of cars is not to exceed 8 pounds per ton; the cylinder power and size of driving wheels are supposed to be properly proportioned to the weight on driving wheels; the track is to be straight and in good order, and the speed no faster than the engine can haul its heaviest loads. The weight on driving wheels is the total on all driving wheels, and the Table applies to 4 or 6 driver locomotives.

PRACTICAL ILLUSTRATIONS OF USE OF THE OPPOSITE TABLE.— Weights on driving wheels are noted at the top of the table, and grades from level to 11 per cent. at the left hand.

EXAMPLE I.—How much can a locomotive with 20,000 lbs. on drivers haul up grades of 4 per 100 ? At the intersection of the 20,000 lb. column and the 4 per cent. grade line is the figure 35, which is the weight in tons of 2,000 lbs. (including cars and loads both) that the locomotive can haul up the grade, and say 18 to 27 tons would be right for daily work, or less for a locomotive with separate tender.

EXAMPLE II.—How much weight on the driving wheels must a locomotive have to haul a train of 40 tons up a grade of 5 per cent ? The number of tons on the 5 per cent. line nearest to 40 is 41 tons, which calls for 32,000 lbs. on the driving wheels ; and for constant work on a long grade, working the engine at about two-thirds to three-fourths of its full power, there should be, say, 40,000 to 46,000 lbs. on the driving wheels.

EXAMPLE III.—If it is desired to haul 50 tons, with a locomotive having 12,000 lbs. on its driving wheels, how steep a grade is possible ? The Table gives the answer, 1¾ per cent., or 92$\frac{4}{5}$ feet per mile, the 50 tons being found at the intersection of the 1¾ per cent. grade line with the 12,000 lbs. column. But for regular work a long grade of about 1¼ per cent. would be the steepest usually advisable.

DIFFERENT METHODS OF DESIGNATING THE SAME GRADES.

Engineer's Method.				English Method.		American R. R. Method.
¼ in 100 or ¼ of 1 per cent.			=	1 in 400	=	13$\frac{2}{10}$ feet per mile
½ in 100 or ½ of 1 "			=	1 in 200	=	26$\frac{4}{10}$ " "
¾ in 100 or ¾ of 1 "			=	1 in 150	=	39$\frac{6}{10}$ ' "
1 in 100	or	1 "	=	1 in 100	=	52$\frac{8}{10}$ " "
1½ in 100	or	1½ "	=	1 in 66$\frac{2}{3}$	=	79$\frac{2}{10}$ " "
2 in 100	or	2 "		1 in 50	=	105$\frac{6}{10}$ " "
2½ in 100	or	2½ "		1 in 40		132 " "
3 in 100	or	3 "	=	1 in 33$\frac{1}{3}$		158$\frac{4}{10}$ " "
3½ in 100	or	3½ "	=	1 in 28$\frac{4}{7}$	=	184$\frac{8}{10}$ " "
4 in 100	or	4 "	=	1 in 25	=	211$\frac{2}{10}$ " "
4½ in 100	or	4½ "	=	1 in 22$\frac{2}{9}$	=	237$\frac{6}{10}$ " "
5 in 100	or	5 . "	=	1 in 20	=	264 " "
5½ in 100	or	5½ "	=	1 in 18$\frac{2}{11}$	=	290$\frac{4}{10}$ " "
6 in 100	or	6 "	=	1 in 16$\frac{2}{3}$	=	316$\frac{8}{10}$ " "
6½ in 100	or	6½ "	=	1 in 15$\frac{5}{13}$	=	343$\frac{2}{10}$ " "
7 in 100	or	7 "	=	1 in 14$\frac{2}{7}$	=	369$\frac{6}{10}$ " "
7½ in 100	or	7½ "	=	1 in 13$\frac{1}{3}$	=	396 " "
8 in 100	or	8 "	=	1 in 12$\frac{1}{2}$	=	422$\frac{4}{10}$ " "
8½ in 100	or	8½ "	=	1 in 11$\frac{13}{17}$	=	448$\frac{8}{10}$ " "
9 in 100	or	9 "	=	1 in 11$\frac{1}{9}$	=	475$\frac{2}{10}$ " "
9½ in 100	or	9½ "	=	1 in 10$\frac{10}{19}$	=	501$\frac{6}{10}$ " "
10 in 100	or	10 "	=	1 in 10	=	528 " "

To reduce grades stated in per cent. (or feet rise per 100 feet of length) to feet per mile, multiply by 52$\frac{8}{10}$.

EXAMPLE.—3 per 100 (or 3%) is equivalent to 3×52$\frac{8}{10}$=158$\frac{4}{10}$ feet per mile.

To reduce grades stated in the English method (or one foot rise in a certain number of feet in length), divide 5,280 by the given number.

EXAMPLE.—A grade of 1 in 20 is equivalent to 5,280 divided by 20=264 feet per mile.

To reduce grades irregularly stated, as for instance, a rise of so many inches in a number of yards or rods or feet to a grade stated in feet per mile, multiply the rise in inches by 5,280, and divide this amount by the length of the grade in inches.

EXAMPLE.—A grade of 5 inches in 1½ rods, multiply 5,280 by 5 =26,400 ; divide by 297 (the number of inches in 1½ rods)=88$\frac{8}{10}$ feet per mile.

EASY METHOD OF MEASURING HEAVY GRADES.

Of course, the proper way of determining grades is by surveyor's instruments. But where the grade varies many times in a distance of a few hundred feet, it is quite as important to know the maximum as the average grade. In such cases it is sufficiently accurate to use a straight edge 100 inches long, and levelling it with an ordinary spirit level, to measure in inches from bottom of straight edge to top of rail. This gives the grade in per cent., which can be reduced to feet per mile by multiplying by 52.8. A few trials in different places will readily determine the ruling grades. On very low grades this method is not practicable, but it is useful on most of the roads where our special service engines are running, the grades varying from 1 to 10 per 100.

CURVES.

THE RESISTANCE OF CURVES is very considerable. The less the radius of the curve, and the greater the length of the curved track occupied by the train, the greater the resistance. The length of wheel-bases of engine and cars, the condition of rolling stock and of the track, and the rate of speed, all influence the resistance, and there is no formula that will apply to all cases.

REDUCTION OF GRADES ON CURVED TRACK. In practice, many engineers compensate for curves on grades at the rate of *two one hundredths of a foot* in each hundred feet for *each degree of curvature*, the grade being stated in feet per hundred.

EXAMPLE.—If a 20-degree curve comes on a grade of five feet per hundred the grade is reduced $20 \times \frac{1}{100} = \frac{1}{5}$ of one foot, which, subtracted from the original grade of 5 feet per 100, leaves $4\frac{1}{5}$ feet per 100 as the compensated grade on the curve ; or, in other words, a grade of 5 feet in the hundred coming on a straight track offers the same resistance as a grade of $4\frac{1}{5}$ feet in the hundred coming on a 20-degree curve.

Where the grade is stated in feet per mile the equivalent reduction for each degree of curvature is $1\frac{56}{1000}$ feet per mile.

EXAMPLE.—A 20-degree curve coming on a grade of 264 feet per mile, the grade is reduced $20 \times 1\frac{56}{1000} = 21\frac{12}{100}$ feet, which subtracted from 264, leaves $242\frac{88}{100}$ feet per mile as the compensated grade on the curved track.

This rule makes no distinction between narrow and wide gauge, and it is doubtful if it applies to very steep grades or very sharp curves. Mr. Nicholas S. Davis informs us that good results were obtained on the Arizona Copper Co.'s railroad of 20 inches gauge, with 4 per 100 grades and 40-degree curves, by compensating at the rate of $\frac{8}{100}$ of a foot per degree of curvature.

In usual railroad practice sharper curves are used on narrow gauge than on wide gauge, because the difference in length of the inner and outer rails on curves on the same degree is not quite so great, and also because the wheel bases of locomotives and of car-trucks are less.

THE GAUGE MUST BE WIDENED ON CURVES. The track should be spread about one-fourth inch on easy curves, and on very short curves about an inch to as much as the tread of wheels will permit. Good results were obtained on the Arizona Copper Co.'s 20-inch gauge railroad by widening the track $\frac{1}{8}$ inch for each 2½ degrees of curvature, making the track on 40-degree curve 21 inches gauge. This is probably about right for most roads using our smaller locomotives and cars of short-wheel base on very sharp curves.

SHARP CURVES UNDESIRABLE. Our smaller special service locomotives on narrow gauge haul mine cars around irregular curves of only 28 feet radius, and they have done daily service around curves of 20 and even 17 feet radius on wide gauge. Our narrow gauge freight and passenger engines are at work on curves of 75 feet radius and upwards; our heavier four driver special service locomotives, on wide gauge, shift cars on curves as short as 70 feet radius. Our larger sizes of motors do good work around curves of less than 60 and 50 feet radius. While our locomotives are capable of conforming to such extremely short curvature, *short or irregular curves are to be avoided, since one bad curve reduces the load that an engine can haul, and bad curves are very destructive to rails and rolling-stock. It is economical to invest more money and get a curve of longer radius, instead of losing continuously in operating expenses.* If, as in the case of mountain and mining roads, a sharp curve is necessary, the rail should be bent to the right curvature. This can be done by a portable rail bender, or by a jack and clamps.

RULES FOR MEASURING THE RADIUS OF A RAILROAD CURVE.

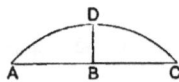

Stretch a string, say 20 feet long, or longer if the curve is not a sharp one, across the curve corresponding to the line from A to C in the diagram. Then measure from B the centre of the line A-C, and at right angles with it, to the rail at D.

Multiply the distance A to B, or one half the length of the string, in inches by itself; measure the distance D to B in inches, and multiply it by itself. Add these two products and divide the sum by twice the distance from B to D, measured exactly in inches and fractional parts of inches. This will give the radius of the curve in inches.

It may be more convenient to use a straight edge instead of a string. Care must be taken to have the ends of the string or straight edge touch the same part of the rail as is taken in measuring the distance from the centre. If the string touches the bottom of the rail flange at each end, and the centre measurement is made to the rail head, the result will not be correct.

In practice it will be found best to make trials on different parts of the curve to allow for irregularities.

Example.—Let A-C be a 20 feet string ; half the distance, or A-B, is then 10 feet, or 120 inches. Suppose B-D is found on measurement to be 3 inches. Then 120 multiplied by 120 is 14,400, and 3 multiplied by 3 is 9 ; 14,400 added to 9 is 14,409, which, divided by twice 3, or 6, equals 2,401½ inches, or 200 feet 1½ inches, which is the radius of the curve.

The formula is thus stated,
$$\frac{A B^2 + B D^2}{2 B D} = R$$

Or applied to the above example,
$$\frac{120^2 + 9}{2 \times 3} = 2,401\tfrac{1}{2} \text{ in.} = 200 \text{ ft. } 1\tfrac{1}{2} \text{ in.}$$

DEGREES OF CURVATURE.

The simplest way of designating railroad curves is by giving the length of the radius (distance from centre to outside of circle) in feet. Civil engineers designate curves by degrees, a one degree (1°) curve having a radius of 5,730 feet, a 2° curve a radius one half as much, a 3° curve one third, and other degrees a proportionate fraction of 5,730 feet, as shown by the following table :

Degrees.	Feet Radius.	Degrees.	Feet Radius.	Degrees.	Feet Radius.
1	5,730	18	318	35	163
2	2,865	19	301	36	159
3	1,910	20	286	37	155
4	1,432	21	273	38	150
5	1,146	22	260	39	147
6	955	23	249	40	143
7	818	24	238	41	139
8	716	25	229	42	136
9	636	26	220	43	133
10	573	27	212	44	130
11	521	28	206	45	127
12	477	29	197	46	125
13	441	30	191	47	122
14	409	31	185	48	119
15	382	32	179	49	117
16	358	33	174	50	114
17	337	34	169		

ELEVATION OF OUTER RAIL ON CURVES.

No rule can be given that will apply to all cases for elevating the outside rail on curves. The gauge of track and kind of traffic, and design of locomotives and cars, all need to be taken into consideration, as well as the rate of speed.

On many standard gauge roads good results have been attained by elevating the outer rail one quarter inch for each degree of curvature. The corresponding elevation for 36 inches gauge would be about one eighth of an inch for each degree of curvature. For the comparatively slow speed at which most of our special service and freight locomotives are generally run, and especially on the extremely sharp curves commonly used, a very much less elevation of the outer rail will be sufficient, and an elevation of 4 to 7 inches for standard gauge, or of 2 to 5 inches for narrow gauge, is probably about the extreme limit needed even on curves of 30 to 80 degrees (or say 200 to 75 feet radius).

RAILS.

We would generally advise for our light locomotives the ordinary T section of steel rail.

VERY LIGHT RAILS NOT ECONOMICAL. The lightest weight of steel rails advisable for the best economy for each size and style of our locomotives is given in the descriptive text with the illustrations. The same weight of iron rails can be used, but not to so good advantage, and steel rails by their greater durability and reduced price have driven iron rails out of the market. It is possible to use lighter rails than we have advised for our locomotives, but it is the best economy to use a rail heavier than is absolutely necessary. Light rails should be made with broad heads as possible, as a very narrow head wears grooves in the driving-wheel tires.

We do not advise strap rails, as they require more expensive track, cost nearly the same as T rails of the same capacity, and are hard to keep in order, and dangerous on account of snake-heads. We have known of light T rails being laid on stringers, and successfully used, instead of heavier T rails on cross-ties. Reversed point spikes are required, and the stringers should be tied across at their top faces by cross pieces let in to prevent rolling or spreading of gauge.

STREET RAILS. For city streets, when T rails are not permitted, probably the best rail is the Johnson rail made with a deep flange.

WOODEN RAILS.

We have built a number of locomotives to run on wood rails, for various gauges from 30 inches to 60 inches, for lumber-mills and other private operations, and also for narrow gauge railroads. We have thus had considerable experience with wooden rails of different patterns and of different kinds of wood. The best wood is maple, laid with the heart up ;

SIZE OF WOOD RAILS. hard pine is used in the South. The simplest form of wooden rails is a stringer cut in 16 to 20 feet lengths, and of such cross section as the kind of wood or weight of engine requires. Five inches square is the size rail we would generally advise, although 5 inches face by 7 inches depth is better. Four inches face by 6 inches depth will answer for our smaller engines, if the wood is good; for large rails 4 feet between centers of cross ties will answer, and for lighter rails 2 to 3 feet between centres. When worn out on top the rail may be reversed, and when again worn out may be used for ties. The ties are easiest fitted and laid if made uniform, and of about the same size lumber as the rails; 6 inches square is heavy enough. Any cheap lumber not especially liable to decay will do. Ordinary hewn ties may be used, but not being uniform are less convenient for cutting out recesses for rails. They should be at least 3 feet longer than the width of the track between rails. The ties must be cut out accurately and uniformly to receive the rails. The recesses should be about 3 inches deep, and be at the top face of the tie one inch, and at the bottom of the recess $1\frac{1}{4}$ inch wider than the rail. The inner faces of the recesses are perpendicular, and the distance between them is the gauge of the track. The bottom of the recess should be level, and ties laid well to afford proper bearing for the stringer.

TIES FOR WOOD RAILS.

WEDGES. Wedges made of any cheap wood, or better, of ends of stuff left from rails, are driven on the outside of rails. They are made of right shape to fit the space left; the reason for making this space wider at the bottom than at the top is to keep the wedges from working up, so that the rail may be held securely in place.

Although our locomotives, especially the designs on pages 20, 39, 35 and 37, are well adapted to wooden rails, we advise steel rails as more desirable and cheaper except in first cost. Wooden rails waste power, are very slippery in wet or freezing weather, require constant repairs, and necessitate very slow speed.

DISADVANTAGES OF WOODEN RAILS.

In some cases it may be best to use them until they earn enough to pay for steel rails, and in the Southern lumber districts where the grades and loads are light and the tracks shifted frequently, it may be well enough to use wooden rails. A light logging locomotive is a very great improvement over animal power whether on steel or wooden rails.

POLE ROADS UNFIT FOR STEAM LOGGING. Pole roads are, in our opinion, unfit for operating by steam. Our experience has been that any one having enough business to justify the use of a locomotive cannot afford to cripple his whole plant for the sake of saving the cost of a track, and that anyone who decides to use a pole road will want a locomotive too cheap to be worth having.

GAUGE OF TRACK.

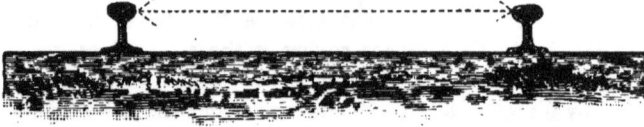

The gauge of a railroad is the distance in the clear between the rails. Our locomotives are built to suit the gauge of track allowing the proper amount of side-play between the wheel-flanges and the rails. A "three foot gauge locomotive" is one adapted to a track with rails just 36 inches apart, and the wheels measure 35¼ inches between flanges. (For the necessity of widening the track on sharp curves see page 54.)

THE GAUGES WE BUILD FOR. We build our locomotives for all gauges of track within reasonable limits, and have built for over 50 different gauges varying from 20 to 72 inches. While we are just as well prepared to build for wide as for narrow gauges, we do not build any but Light Locomotives and our largest cylinders are 14 inches diameter. We have built locomotives with 9½ inches diameter cylinders for 20 inches gauge, 12 inches cylinders for 30 inches gauge, and 7 inches cylinders for 72 inches gauge.

COST OF NARROW GAUGE AND OF WIDE GAUGE LOCOMOTIVES. Correspondents frequently request "prices for both narrow and wide gauge engines," and sometimes for 24, 30, 36, 48 and 56½ inches gauge, under the impression that the narrower the gauge, the cheaper the locomotive. A very wide gauge is undesirable for a very small locomotive, and an extremely narrow gauge involves modifications in design which increase the cost of all but our smallest sizes ; but except for such extreme cases *there is no difference in price between a wide gauge and a narrow gauge locomotive* of the same design and same size of cylinders.

SPECIAL GAUGES USED FOR SPECIAL PURPOSES. The metre gauge (39⅜ inches) is common in foreign countries. There are a number of roads at home and abroad of 42 inches gauge. Plantation tramways in Spanish countries and steel mill tracks in this country are often 30 inches gauge. For copper, silver and other mines 20 to 30 inches gauge is often adapted to save cost in under-ground work, and similarly narrow gauges are often desirable throughout the yards and buildings of manufactories. Many operators of bituminous coal mines prefer a gauge of 40 to 44 inches, because it admits a desirable shape and capacity of mine cars. Street railways are quite commonly 60 or 62½ inches gauge and no change of gauge is needed when animal power is abandoned for our Steam Motors. Odd gauges of track are frequent for private and local roads, because some whim or trivial reason determined the gauge at the start. In some instances the saving of a few dollars in buying second-hand equipment of odd gauge has resulted in an extensive system of odd gauge railway and a very great subsequent outlay to change the gauge.

THE 'NARROWER GAUGE' AND "NO GAUGE" SYSTEMS. The "narrower gauge" of 24 inches has been recommended as the best gauge for short roads for freight and passenger traffic and is entirely practicable. But except for some mill or mine tracks the 24 inches gauge has no advantages and has some disadvantages as compared with the 36 inches gauge. There is no saving in cost of construction or operation, no gain in efficiency, and the power and the variety of design practicable for locomotives are limited by very narrow gauges. There are also various "no-gauge," "peg leg," "saddle-bag" and similar systems requiring still greater modifications and without any recommendations that we know of unless their novelty will induce curious people to invest in them.

ODD GAUGES UNDESIRABLE. While our locomotives for all gauges of track are thoroughly efficient, and we have overcome all mechanical difficulties in adapting them to very narrow gauges, very wide gauges, and all odd intermediate gauges, we believe, unless there are exceptional reasons to the contrary, that our customers in planning new roads will serve their own interests best by adopting either the regular narrow gauge of 36 inches or the standard wide gauge of 56½ or 57 inches. Equipments of odd gauge cannot be obtained or disposed of promptly.

COMPARATIVE MERITS OF NARROW AND OF WIDE GAUGE.

The principal advantages claimed for narrow gauges are adaptation for sharp curves and steep grades, lighter rails and equipment, and cheapness in cost of construction, also better proportion of paying load, less wear on rolling stock, and cheapness of operating.

EFFECT OF GAUGE ON CURVES AND GRADES. Two surveys are often made for a proposed road, one for an expensive wide gauge with heavy rails and rolling stock over easy grades and curves, and the other for a cheap narrow gauge with light rails and rolling stock over steep grades and sharp curves. Over very mountainous country with heavy cuts and fills, and especially with a great amount of hill-side work, the excess of cost of grading due to the difference in gauge of track may be an important item. But over ordinary country the same grades and curves, and rails and equipment of the same weight may be used for the wide gauge. Our estimates of cost per mile of track on pages 83 and 84 apply to either gauge.

The narrow gauge admits sharper curves, because the wider the gauge the greater the amount of slipping of wheels in passing curves ; but practically this is too small to consider unless on curves too sharp to be desirable on either gauge for ordinary purposes. Sharper curves are commoner on narrow gauge because smaller locomotives are generally used.

The resistance of gravity and the power of a locomotive on grades are just the same, no matter what the gauge of track may be, but some features of usual practice make a slight difference. The wide gauge increases the weight enough to be appreciable in the case of very small locomotives ; short wheel-bases on wide gauge have more tendency to crowd against the rail ; a train made up of a few large wide gauge cars has less friction and may be easier to haul than a narrow gauge train of the same weight made up of a larger number of lighter cars, but the narrow gauge train is easier to start by taking up the slack. Our figures of hauling capacity apply equally well to all gauges, and other conditions than gauge of track will determine in each case the most convenient loads for daily work.

GAUGE OF TRACK AND PAYING LOADS. When the narrow gauge system was first agitated it was argued that wide gauge cars could not be built as light as narrow gauge and carry the same load. Wide gauge cars have since been re-modeled so that in actual practice there is no marked difference between the two gauges in the proportion of dead to paying weight.

The principal objections urged against narrow gauge are : top-heavy rolling stock with limited speed and power ; and transfer of freight and passengers.

GAUGE AND SPEED AND POWER. Our narrow-gauge locomotives, both with tender and with saddle-tank, are not in the least top-heavy, and have frequently attained speeds of 30, 40, and even nearly 60 miles per hour. If more power is needed than about 13 inches diameter of cylinder, the wide gauge is preferable, though not necessary.

BREAKING GAUGE. Transfer of freight and passengers may in some cases be unobjectionable, and may be desirable even when not made necessary by difference in gauge. There are a number of successful systems for transfer of freight without breaking bulk. But the need of interchange of cars, and the advantage of a uniform gauge, have led to the widening of many narrow-gauge roads, both "feeder" lines and competing lines, even where the traffic was easily within the capacity of the narrow gauge.

The question of gauge of track is of much less practical importance than the question of

LIGHT RAILROADS.

Our locomotives are the best motive power for a very great variety of roads where a heavy expensive road would be impracticable, mechanically or financially, and where reliable service is desired at a moderate cost of construction and operation. When the work to be done is within the limits of a 16 to 25-lb. rail the narrow gauge may often be preferable, as in the case of many contractor's tracks ; plantation, coal and ore roads ; and some logging roads and light motor lines. When anything heavier than a 30-lb. rail is needed, as may often be the case with contractor's, logging, suburban and motor roads, the standard gauge is usually more desirable. For a very large proportion of roads for which our light locomotives are used, there is but little choice between narrow and wide gauge except as special conditions may exist in each case. All the advantages of the narrow gauge system are also secured by light railroads of standard gauge, but when connection is made with trunk lines a 30-lb. rail is necessary to carry the cars, and usually nothing smaller than a 10 by 16 cylinders locomotive is advisable.

STREET RAILWAYS, AND RAPID TRANSIT AND "DUMMY" MOTOR LINES.

We offer our noiseless Steam Motors, described on pages 32, 33, 42, 43, 44 and 45, as affording, in great variety of size and design, the least expensive and most desirable motive power, both as a substitute for animal power on city streets and for many local passenger purposes for which animal power is wholly inadequate.

Our motors are simple and durable in construction, and without objectionable or complicated devices. The general design and quality of work and material are in no respect inferior to the best locomotive prac-

DETAILS OF CONSTRUCTION OF OUR MOTORS. tice, no cogs, gears, upright boilers, or gas pipes for conveying steam being used. The patent noiseless exhaust used is effective and durable and placed where it is not in the way or liable to be injured or get out of order ; it converts the usual intermittent noisy action of the steam into a continuous, quiet flow, without back pressure. The expensive, cumbersome condensing arrangement used on foreign "tram way's engines" is found unnecessary in our own more practical country, as with the patent exhaust, no steam is noticeable under ordinary working conditions. Smoke is avoided by the use of anthracite coal or coke fuel. About 8 to 12 pounds of anthracite coal per mile is usually sufficient, although in some cases with heavy loads and steep grades, 15 to 20 pounds per mile is used, and very much depends on the engineer. Crude petroleum fuel can be used with special appliances, but in addition to mechanical difficulties it is too expensive. The machinery of our motors is enclosed in a cab so that they resemble horse-cars or railway-cars so nearly that no difference is detected at the first glance. The motor cabs are substantially built and handsomely finished, and roomy and conveniently designed ; glass sash is arranged to drop all around, and at the front end reaches to the floor ; hinged trap doors in the cab floor give opportunity for oiling the machinery in motion ; and the fuel bunker is of ample capacity and handily placed. In all our motors the engineer has a good look out and full control of all valves and levers so that the motor can be stopped or started instantly.

BEST DE-SIGNS FOR CITY STREETS AND SLOW SPEED AND STEEP GRADE. The motors without pony trucks, described on pages 32 and 33, are best adapted to slow speed, as is usual where the road is wholly on city streets. The smaller sizes, say 7 x 12 and 8 x 14 cylinders, are ample for hauling on ordinary grades one to four cars : and the larger sizes are desirable for hauling a number of cars up steep grades. The rear-tank design described on page 32 has the dome, engineer's seat, valves, levers, etc., placed centrally and gives the most perfect outlook in all directions. The saddle-tank design, page 33, more nearly resembles a street-car, and permits the shortest possible length over all, and the position of the tank over the boiler does not interfere to any objectionable extent with the engineer's outlook, except for the largest sizes for which a fireman would generally be required.

BEST DE-SIGNS FOR POWER AND SPEED COMBINED. The motors with back-truck, described on pages 42, 43 and 44, are best adapted to work requiring a combination of speed and power. The small sizes are useful for hauling a limited number of street-cars where, for part of the way at least, there is an opportunity for considerable speed, and the larger sizes are desirable for suburban roads, hauling longer trains and heavier cars. The designs described on pages 42 and 43 carry the water in a saddle-tank over the driving wheels, and thus have the greatest power that can be secured in combination with the easy motion and speed afforded by the pony truck. This position of the tank is not objectionable in the smaller sizes, but interferes with the engineer's outlook for the larger sizes enough to make a fireman desirable. The design described on page 44 gives a perfect outlook in all directions, with a dome, engineer's seat, levers and valves placed centrally, with a very roomy, conveniently arranged cab, and is the most popular style. It is not quite so powerful and on extremely steep grades not so desirable as a saddle-tank motor.

The motor described on page 45 has a pony truck at each end which makes a saddle-tank necessary to get sufficient weight on the driving wheels. It is not the best design for very heavy loads and very steep grades, but is the fastest possible motor, and very well liked by roads using it.

SHARP CURVES. For all of our motors with pony trucks we use a special patented truck which enables them to pass curves very easily, and to work constantly on curves that most railroad engineers would pronounce impracticable. Our 12 x 18 cylinders motors with back-truck are at work on quarter circles of considerably less than 50 feet radius.

Our Motors are constructed to run equally well in either direction, and with entire control and good outlook by the engineer running forward or backward.

BEST RAIL. The best rail for our motors is a steel T rail of suitable weight, as this allows the usual depth of wheel flange and width of wheel tread, and dirt and stones cannot rest upon it. When city ordinances forbid a T rail the best rail is the Johnson street-rail, and the deeper patterns are preferable. We make the tires of our motors to suit any special rail that may be used.

COMBINED MOTORS. Various "combined" motors and cars, in which the car and engine is contained in the same machine, have been tried but have proved deficient, and are now almost out of date, and superseded by the separate motor. The combined car and motor has the merit of taking up the least possible room. But this arrangement cramps the machinery, compels the objectionable vertical boiler and a wheel base too long for ordinary street curves, makes the car too rough to ride in or else too shaky for the machinery, and annoys passengers with the vibratory motion of the engine, and the heat of the boiler and the smell of oil. Thus the car and engine are both spoiled, and, in addition, any repairs to either lays both up.

COMPRESSED AIR, SODA, FIRELESS AND OTHER MOTORS. Various machines operated by compressed air, or by ammonia and other volatile chemicals; also steam motors, condensing and using the steam over again, or arranged for charging with fresh steam, or for renewing the steam by hot soda reservoirs; also coiled spring motors, thermo-motors, and many other ingenious contrivances have been invented, and announced as the coming motor about to revolutionize railroads, and then have been abandoned as failures. Thus far only two adaptations of mechanical power for street railroads have any real claim to be considered rivals of steam motors, viz., electric systems and cable systems.

ELECTRIC MOTORS. The latest and, perhaps, the most popular substitute for direct steam power is electricity. There are a great many systems of electric railroads with overhead "trollies" and dangling wires, or with "conduits" for underground wires; also storage batteries carried on the motor. These roads have proved the mechanical possibility of hauling street cars up very steep grades and around sharp curves, and at a good rate of speed by electricity, but have at the same time made evident the great and, perhaps, insurmountable difficulties of satisfactory and economical continuous operation. The storage battery seems to be the most desirable electric system, because it avoids obstructions in the streets and dangerous naked wires; but even when enough battery power is used to make the weight undesirable the power is limited, except at the risk of its speedy destruction. Until some absolutely new discovery, the expense of the storage battery makes it only an interesting experiment without any commercial or practical utility. Except in a few cases where water power is utilized, electricity for street use must cost more than the direct application of steam.

The reasons for this are the excessive cost of maintenance and of interest on the permanent plant, and also the immense waste inseparable from every conversion of power into electricity and back again into power; because, whatever else electricity may be, it is not power, but only a means of transmission of power. Financial reasons are in our opinion decisive against electricity, but, in addition, is the more important matter of danger to life and property. Almost every American street is already encumbered with a network of wires for telegraph, telephone, fire-alarm, or police-patrol purposes, and for electric lighting. The naked wires used for every practicable electric motor system may at any moment, by mere contact with any other wire or conductor, divert a current fatal to life and destructive to property. On account of frequent groundings and other mishaps peculiar to electricity, travel by electric motors is liable to indefinite stoppage at any time without notice, and there is already some demand for steam motors as a reliable reserve power for electric roads. Unless apparent impossibilities are accomplished, we believe that electric motors, which, because of the popular demand for novelty and readiness to believe anything not understood, are often easy to introduce, will, by calling attention to the need of some cheap and reliable power, increase the sale of steam motors.

The cable road is the only system which in any considerable number of cases is preferable to steam motors. Its positive application of power **CABLE ROADS.** saves the room needed by any separate motor depending on rail adhesion, and also is adequate for a very heavy business, and inclines impracticable for other systems can be ascended at fast speed, and any extra rush of travel can be accommodated promptly by merely attaching more cars to the "grip" car. If the business is large enough and the distance not too long, these advantages may overbalance the immense cost of the cable system, the astonishing waste of power, the rapid wear of the cable, the danger of accidents, the damage to the streets by slot-rails and manholes, and the stoppage of the whole line inevitable in case of accidents or repairs for any part of the line. Our steam motors are valuable to cable roads for use on extensions and also as a reserve ready to use in case of need.

DECISIVE ADVANTAGES OF STEAM MOTORS. The separate steam motor is not only the least objectionable, most serviceable and least expensive system for street railroads, but in one most important respect it differs from all other systems and is preferable to them. There is no outlay for any battery of stationary boilers, engines, power house, dynamos, compressors, overhead poles, wires, underground conduits, cables, man-holes, slot rails, torn-up streets, and no interference with telephone, electric light, and telegraph wires, sewers, gas and water pipes, etc. The steam motor only needs to be fired up and run, and this can be done without interrupting horse-car service. The "experiment" only involves the difference between the cost price of one motor and what it can be sold for as a second-hand machine, instead of many thousands or hundreds of thousands of dollars.

THE ONLY REAL RIVAL OF THE STEAM MOTOR. The principal obstacle, and in many cases a sufficient one to the use of the steam motor on city streets, is one which applies in greater measure to other applications of mechanical power. Where horse-power in crowded streets is fast enough the greater speed of steam cannot be used; and when horse-power is able to haul any loads to be hauled the greater power of steam is of no advantage. In such cases, although the steam motor is more economical and the outlay but little greater, conservatism will adhere to old and well-tried methods and be slow to abandon horse-power.

PRINCIPAL OBJECTION TO STEAM MOTORS. As compared with other systems the steam motor makes no more noise than most horse cars, cable roads or electric motors, and in general appearance is no more liable to objection. The principal obstacle to be overcome is popular prejudice, and the best way to overcome this is to run a steam motor a short time.

SUBURBAN RAILWAYS AND MOTOR LINES; HOTEL AND EXCURSION ROADS.

STEAM MOTORS WITHOUT COMPETI-TION FOR SUBURBAN SERVICE. The movement of city populations toward their suburbs is in an increasing ratio every year. Horse-power is too tedious, fails to meet the requirements, and is too expensive. Any advantages that cable or electric roads are supposed to have for short runs disappear as the length of the road is increased. The utility of the steam motor is on the contrary more evident as the run grows longer. When city ordinances and ignorance prevent the use of steam motors on the city streets we advise their use for the out-of-town part of the run. Our motors are useful and money-making on extensions of electric or cable roads where the great expense of these systems and the amount of business offered would not justify these systems. Land companies may, by building a "rapid transit" line at a moderate cost, put their property on **MOTOR LINES AND REAL ESTATE.** the market at great profit, and have, besides, a good paying investment in the road. Proprietors of summer resorts, watering-places, hotels, excursion and picnic grounds, often find their business limited by the difficulty of transporting any large number of people in a short time.

This difficulty can be solved satisfactorily by a light-equipped "dummy" line. In most cases where the season is short or the business irregular, sometimes very light and sometimes very great, the best economy is to lay a rather light rail, and not to use motors excessively large but of medium power, and to have one or two motors and a proportionate number of cars in reserve for special occasions. Roads of this character are not only profitable to the owners, but are also a great public benefit. Suburban roads need not cost, exclusive of franchise, land and buildings, over $3,000.00 to $6,000.00 per mile, and will earn as much as the suburban trains of existing main lines which have cost five or ten times as much. Even if suburban roads made no profits, they would often be worth their cost by securing rates and facilities independent of foreign or hostile managements. The gauge of track may be 36 or $56\frac{1}{2}$ inches as circumstances may make the more desirable. For a great number of purposes unenclosed motors are more desirable than enclosed motors, and we are prepared to substitute cabs similar to those of ordinary locomotives, and at a considerable reduction in price.

Motor roads are often built by men who are not personally familiar with the details of railroad machinery and management, but who can see that such roads are paying investments. We wish to urge upon capitalists and organizers of new motor lines the necessity of having not only good motors, good cars and good track, but also of having some

A MONEY SAVING HINT. competent, experienced railroad man, who will know how to keep everything in running order. The lack of such a man may mean failure and is sure to involve a loss of more money than his salary would amount to. On small roads running but one or two motors

he may also serve as engineer. On motor roads the service is always severe ; mud, dust, sharp curves, uneven grades and constant stopping and starting demand good care of machinery ; small engines on short runs with frequent stops are expected to make a greater mileage than is made by large locomotives on long roads. It is very short-sighted policy for a motor road, after demanding and getting the very handsomest and most efficient machinery with all the latest improved appliances, to let their motors and cars lie out in the weather without protection or care. It is a very costly economy to hire the cheapest engineers, or to let the track get out of line and sunk into the mud or to jump trains over rails at crossings.

SPECIAL SERVICE.

FURNACE AND CINDER LOCOMOTIVES. Iron furnaces are usually so located that fuel, limestone and ore, or metal or cinder, must be moved to and from different parts of the works. Here the cost of wagon-hauling on dirt roads is so excessive, that a rail track, either wide or narrow gauge, as may be most

convenient, is essential to economy and successful competition.

the work is more than three animals and drivers can do (see pages 77 and 78), a special service locomotive (see pages 26, 34, 38, 22, 40 and 41) is required, and will very soon pay for itself.

At Bessemer steel works these special service locomotives are used for hauling hot ingots from the converter, and have proved so useful that

This cut shows one of our 9 x 14 cylinders locomotives on the cinder bank of the Chestnut Hill Iron Ore Co., Columbia, Pa., and unloading cinder cars by the patent steam attachment of Mr. Jerome L. Boyer, Reading, Pa. (See page 125 for working report and description.)

they are now an established part of the plant. It is also practicable to haul molten metal a distance of several miles from blast furnaces to the converting-house, instead of casting and re-melting **STEEL-WORKS** the pig iron. Our smaller special service locomotives **LOCOMOTIVES.** are also useful in hauling hot blooms to the rolls in rail-mills and other large steel-mills. For hauling ingots, hot metal, cinder, etc., the locomotive cabs and other parts usually of wood are made of iron to endure the exposure to the intense heat. When the locomotive works inside of the mill under cover the cab may be omitted and a long coupling bar used. (See pages 40 and 41). It is well to select a larger locomotive than absolutely necessary for hauling hot loads, as the cars are heavy and clumsy, and the oil often burned off of the car journals.

Our locomotives are used for handling fluid metal, ingots, blooms, etc., through the following large steel works : North Chicago Rolling Mill, Union Iron and Steel Co., Joliet Steel Co., St. Louis Ore and Steel Co., Pennsylvania Steel Co., Scranton Steel Co., North Branch Steel Co.,

Midvale Steel Works, Otis Steel Works, The Edgar Thomson Works, and Homestead Works of Messrs. Carnegie, Phipps & Co., Ltd., Linden Steel Co., Jones & Laughlins, Ltd., Pittsburgh Steel Casting Co., Messrs. Miller, Metcalf & Parkin, Messrs. Oliver Brothers & Phillips, and at over fifty iron-mills and blast furnaces.

Many large manufacturing establishments have found it the best economy to use our special service locomotives for moving raw and finished material through their works. When a track of 24 to 36 inches gauge is used for connecting the different departments, our smaller sizes of special service locomotives described on pages 26, 38, and 34 are oftenest used. When a standard gauge track is adopted, and usual freight cars moved, larger locomotives are desirable, either the larger sizes of pages 26, 38 and 34 or some of the sizes on page 24. These locomotives are used at copper and silver smelting works, iron, gold, silver, copper, fire-clay and phosphate mines; cement, lime and building-stone quarries; at brick-yards, and at manufactories of cars, car-wheels, tires, plate-glass, sewing-machines, mowing and reaping machines, threshing machines, wooden ware, etc., and are adapted to many other purposes, of which no detailed account can be given.

GREAT VARIETY OF MANUFACTURERS USING OUR LOCOMOTIVES.

RAILROAD SHIFTING.

Engines unnecessarily heavy are often used for shifting where our larger sizes, described on pages 24, 21, or 23, would do the work as well, and at less cost. These engines are very compact and powerful, start their trains quickly, and work on steeper grades and sharper curves than ordinary railroad shifting engines. The process of shifting cars by animals, or by a gang of men with pinch bars, is a most inconvenient extravagance, as is also any dependence on railroad companies for occasional use of shifting engines. In such cases, without counting the gain in time, comfort and convenience, it does not take long for our locomotives to save their cost.

ECONOMY OF OUR SHIFTING ENGINES.

CONTRACTOR'S WORK.

Contractors who have any considerable quantity of rock, mud, or ◆ earth to move, can do it most economically by our special service locomotives, such as are described on pages 20 to 26 and 34 and 38.

The gauge of track for contractor's tram-ways may be narrow or wide as most convenient. Narrow gauge is best where the plant needs to be shifted often, and some contractors prefer 30 or even 24 inches gauge for this reason, and use very small locomotives and cars. Usually there is no advantage in anything narrower than 36 inches gauge. When standard gauge cars belonging to the railroad can be used for grading to good advantage, 10 x 16 cylinders is usually the smallest size locomotive desirable. Often when narrow gauge is used, the heavy rails intended for the finished railroad may be used instead of lighter rails.

ECONOMY OF CONTRACTOR'S LOCOMOTIVES. One of our contractor's locomotives, or two if the haul is long or grades steep, will keep a steam shovel busy. It pays to use a locomotive even for hauls as short as 500 or 1000 feet. Compared with animal power, our locomotives save their cost many times over ; compared with other locomotives, they are efficient and durable and will stand hard usage 24 hours per day constant use six days per week with reasonable care. In case of accidents our locomotives are only laid up, if at all, long enough for a telegram to reach our shops and supplies expressed to reach destination.

Our contractor's locomotives have proved useful in the construction of the following large works: The United States Government Works at Muscle Shoals, Yaquina Bay, Columbia River Cascade Locks, and the Mississippi Rapids near Keokuk ; The Panama Canal; the Hoosac, Musconetcong, Pittsburgh Junction, Hoboken, Baltimore and other tunnels; the Northern Pacific Railroad, both in the laying of the first track and in the completion of the Cascade Tunnel; the Montclair Railway; Canada Southern R. R. ; West Shore R. R. ; South Pennsylvania R. R.; Illinois Central New Line ; the improvement of the Pennsylvania, Baltimore & Ohio, and Shore Line railroads ; the deepening (and subsequently the filling) of the Providence Cove, the filling of the South Boston Flats and of the Potomac Flats ; the Hiland Reservoir at Pittsburgh, the new Reservoir at Washington, and the Croton Aqueduct. Reports of the workings of some of these locomotives may be found on pages 110 to 125.

COAL ROADS.

When coal is sent to market by water it is generally best to run the mine cars to the water, and sort and ship the coal in one operation. Where coal is shipped by rail it is usually cheaper to extend the mine road than to build a branch of the wide-gauge road several miles to the mine. The excess of the cost of the wide gauge over the narrow gauge, on which the mine cars are hauled by a light locomotive, like those shown on pages 22 or 26, would often be enough to pay for the entire rolling-

This cut shows the Tipple for shipping bituminous coal by river. The coal is hauled from the foot of the incline or from the mine by a locomotive and is dumped into flat boats. The nut coal and lump coal are separated by screens and loaded and weighed into different boats. When coal is shipped by rail the flat cars are loaded by a similar arrangement.

stock of the latter. The best results are obtained when loaded cars go down and empty cars go up grade. When the locomotive has brought its loaded train to the tipple or breaker, it should find an empty train ready, and when this empty train has been brought back to the mine it should be exchanged for another loaded train without delay. At each terminus there should be two tracks, one for empty and one for loaded trains, and the grade should be so adjusted that the cars may be handled by gravity. The exercise of a little foresight in the location and details of such a road, with reference to economy of handling and shipping, may, with little or no addition to the outlay, save a large amount every year.

COAL MINES.

In adapting our locomotives to inside use in mines difficulties were encountered and overcome. The grades and curves are usually excessive, and the rails light and often wet; considerable power is required in a very contracted space; dry steam must be obtained with low steam-room; even where the head-room is not enough for a man to stand upright, the locomotive must be provided with a comfortable place for the engineer, with everything placed conveniently within his reach and control.

The dimensions of openings and weights of rail required for different sizes and styles of mine locomotives are given, with the illustrations and descriptive text on pages 28, and 30. We advise the larger openings as giving the best and most economical results.

In hauling under ground, as in outside hauling, animals cannot compete with locomotives in economy and efficiency. The table of comparative cost is given on page 77.

The principal objection against mine locomotives is, that the smoke is injurious to the miners. Its best answer is an actual test properly made. Experience makes mine locomotives popular with miners, since, if annoyance is felt from the smoke, the ventilation of the mine is shown to be defective, and the mine operator, to secure to himself the advantages and saving obtained by the use of the locomotive, must secure to the miners a proper supply of pure air. Thus the locomotive not only has done no harm, but has pointed out an existing danger, which was the more hurtful because imperceptible. Bituminous coal is better than anthracite, and coke is worse than either. Even where mines are badly ventilated a mine locomotive does good, rather than harm, since by its passage through the entry, a draught is made, which expels the foul air and smoke together. It is only necessary to supply the mine rooms with fresh air independently of the main entry, which is the best and simplest method of ventilation, whether a locomotive is used or not. As no two mines are exactly alike the arrangements of details of ventilation will vary; but the one thing essential is to use the entry where the locomotive works for the out-current of air and not for the in-current. A furnace or a fan may be used as may be most convenient. For tunnels open at each end natural ventilation is usually sufficient.

Our mine locomotives are in use in the anthracite and the bituminous regions of Pennsylvania, Maryland, West Virginia, Virginia, Ohio, Kentucky, Georgia, Tennessee, Illinois, Iowa and Washington Territory. Some of them have been in constant use for ten years two and a half miles underground and very seldom coming out into daylight.

Reports of some of our mine locomotives are given on pages 126 to 131, and the different sizes and designs are described on pages 28, 30, and 31.

COKE OVENS.

The manufacture of coke from bituminous coal for use in blast furnaces, iron and steel mills, and also in the form of crushed coke for use in dwellings, has developed so that the ovens can no longer be charged in the old-fashioned way by cars drawn by mules. Our light locomotives described on page 41 are especially constructed for this work, having sheet-iron cabs for protecting the engineer, and they haul one to five larries at a trip, charging 100 to 300 ovens per day, according to the size of the locomotive and the grades and distance. The gauge of track is usually 56½ inches, and very sharp curves are often necessary. The double-row system of ovens is the most convenient, with the track laid between the ovens and with larries with a spout on each side ; but the old system with the track over the centre of the ovens can be used. It is cheapest to use a heavy rail of 50 to 60 pounds per yard, bearing on pillars, and not to have the weight of the locomotive and larries rest on the ovens. When heavy rails are used the driving wheels of the locomotive may be solid chilled iron, which are cheaper than steel-tired wheels, and do not require turning down, and for these reasons may be preferable.

A locomotive with 7 x 12 cylinders is generally amply powerful for coke-oven service, and often a 6 x 10 cylinders locomotive is sufficient. The locomotive may also be utilized for shifting the usual railroad cars for loading. In some cases it may be desirable to use the same gauge of track on the ovens as for the mine cars and haul the mine cars as well as the larries.

A few reports of coke-oven locomotives are given on pages 110 to 125.

LOGGING RAILROADS.

Steam railroads with proper locomotive and cars, furnish the cheapest and most reliable plan for moving logs from a timber track to the water. They are equally desirable in many cases for hauling logs to the mill or to a main line of railroad.

The best gauge for most logging roads is 56½ inches, because wide gauge cars can have extra long bolsters and be loaded heavily without piling the logs high. For light logging roads with rails of 16 to 20 lbs. per yard, the narrow gauge of 36 inches may be preferable. Odd gauges are to be avoided, as their rolling-stock cannot be bought or disposed of to as good advantage as for regular gauges.

This cut represents a 7 by 12 cylinders locomotive hauling 17,650 feet of logs on 10 cars, 8 miles in 33 minutes, on a 20 lb. per yard iron rail.

The best rail is steel, of 16 to 40 pounds per yard weight, according to the work to be done. Instead of earthwork fills or trestles, imperfect and unmarketable logs may be built into cribwork for crossing swamps and other depressions. The rails are then laid on stringers, and reverse point spikes are used ; the stringers are tied across at their top faces to prevent their rolling, as explained on page 56. Our experience with wooden rails is also given on pages 56 and 57.

A logging road should be equipped with enough cars for two trains, one to be loading while the other is on the road, so that the locomotive need not wait for cars to be loaded. The unloading can be done so quickly as to cause no delay.

Our locomotives are well adapted to this service. Those described on pages 26, 34, and 38 are often used, as they are the simplest and least expensive. The back-truck styles on pages 20, 21 and 39 are generally most desirable as they can make the greatest number of trips and also haul heavy loads. Pages 22 and 23 are preferable for excessively steep grades where power rather than speed is required. Pages 8, 12, 16 and 36 are desirable for extra long runs.

Logging railroads are generally so built that the service is very severe, and there are few places where it is so poor economy to use cheaply-constructed locomotives. A large force of men and an expensive invest-

ment may be rendered useless by the attempt to save a few hundred dollars in motive-power. Good mules are preferable to poor steam machines.

COST OF HAULING LOGS. The cost of hauling logs by our locomotives, including interest and depreciation, and all expenses, varies from about 30 cents to 60 cents per 1,000 feet, according to the length and general condition of the road, and the amount of business. The cost of hauling by horses with sleds over snow, or iced tracks, is usually $1 to $2.50 per 1,000 feet, allowing two to three trips per day. A lumberman dependent on sledding is liable to have his operations entirely suspended by a mild winter, and his money locked up for a year at least. Meantime, his logs are depreciating in value, and are unsalable when prices are the highest and the demand greatest. By building and operating a logging railroad, however, he may still reach the season's market, and afterwards carry logs all the year round. When prices are high the output can be doubled, without additional investment, by running 24 hours per day ; or, on the other hand, when prices are low, and operations therefore suspended, all expenses are stopped. When timber has been injured by fire or windfall, it may be brought to market before it can be destroyed by decay or boring worms by building a logging railroad. The entire outlay for a steam logging road with steel rails is about 50 cents or $1 for each 1,000 feet of lumber readily reached by it. When the tract is cut off, the road may be moved to another tract at slight expense. Under reasonably favorable conditions a logging railroad more than pays for itself inside of a year. The investment is a paying one, even if the timber reached is cut off, and the road moved to open up another tract every year. Tracts, before considered of little value and inaccessible, may be utilized and worked to make even more profitable returns in proportion to the investment than lands held at a higher figure because more favorably located. Logging railroads solve the problem also of the economical and profitable production of lumber, where otherwise the cost of moving, as it increases with the length of the haul, leaves after each year's cut a diminishing margin of profit. This low cost of transportation enables "culled" or poorer grades of logs—which by any other method of logging would be left to rot in the woods—to be marketed with profit, and logs can be sold with a handsome margin at what are cost figures to operators hauling by animals.

The advantages and economy of logging locomotives are by no means confined to immense operations. While our larger locomotives can put in 1,000,000 feet per week on a haul of 5 to 10 miles, our smaller locomotives are just as economical and almost as indispensable for any mill cutting say 15,000 to 20,000 feet daily and hauling logs or lumber over a half mile.

Our locomotives are hauling logs in Pennsylvania, the Southern Atlantic and Gulf States, the Northern Lake States, and on the Pacific coast. The total extent of territory annually denuded of timber hauled by locomotives built by us is about 350 square miles.

Our locomotives are also used for sorting and piling lumber in lumber yards, and for hauling sawdust and waste from the mill to a refuse burner.

WORKING REPORTS are given on pages 132 to 147.

PLANTATION RAILROADS.

In the West Indies, Mexico, Sandwich Islands, South America, and in our own Southern States, our light locomotives are used on plantations for carrying sugar-cane from the fields to the crushing-mill, and for shipping sugar and molasses, and for receiving fuel and other supplies. The gauge of track is usually 30 or 36 inches, and the metre gauge is sometimes used.

The service is peculiarly difficult in several respects, and demands locomotives well adapted to the requirements. The soil is usually very soft, and in the rainy season the rails are sometimes hidden by the mud ; a light or portable track is often used for convenience in moving the road in the fields ; the road follows the contour of the surface of the country, and the curves and grades are frequently excessive ; the climate is very hot and moist, and good engineers are not alway obtainable. The Plantation Locomotives on pages 34, 35, 37 and 14 meet all these conflicting conditions, as they are light, compact and powerful, and with their weight well distributed ; the different parts are strongly made to stand rough usage, and the cabs are open to secure the comfort of the engineer. If desired, greater power may be gained by carrying the water over the boiler (as shown on pages 20, 21, 22, 23, 26, 38 and 39), but plantation owners generally prefer the rear tanks. Wood, coal, gas-house coke, or the refuse dry-pressed cane, may be used as fuel.

Plantation locomotives are applicable to any large farming operations, and, with such modifications as the climate and the conditions of the service may require, are just as capable of saving time and money in the great wheat-fields of the Northwest as in the plantations of the tropics.

COMPARATIVE COST OF OPERATING ANIMALS AND LIGHT LOCOMOTIVES.

The following calculations demonstrate that on an average where three animals and three drivers, or animals and drivers in different proportion, but at about the same daily expense, are used, it is cheaper to operate a light locomotive. From $5 to $6 per day, or $1,500 to $1,800 per year, is a reasonable allowance for the cost of operating a light locomotive, to take the place of 10 to 30 animals. It is not unusual for an engine to save its cost in less than a year. When, through strikes or dulness of trade, an engine is idle, it saves money as well as when it is busy ; only a few cents of white lead and tallow are needed for it, while mules, whether idle or not, must be fed.

Cost per year of operating 3 mules and 3 drivers.

Where Feed and Labor are at	Low Prices.	Average Prices.	High Prices.
3 mules' feed, harness, shoeing, care, etc., for 365 days, each per day.....................	@33⅙c.=$365.00	@ 60c.— $657.00	@$1.00—$1,095.00
3 drivers' wages, 300 days, each per day.....................	@75c. — 675.00	@ $1.25—1,125.00	@ 1.75— 1,575.00
8 per cent. interest, mules worth $150 each........... .	— 36.00	— 36.00	— 36.00
Total..........................	$1,076.00	$1,818.00	$2,706.00

Cost per year of operating one of our light locomotives, capable of doing the work of 10 to 30 mules or horses.

Where Fuel and Labor are at	Low Prices.	Average Prices.	High Prices.
Oil and repairs, per year......	$30.00	$100.00	$200.00
Fuel, 400 to 1,000 pounds coal, or ⅛ to ¾ cord wood. Costs almost nothing at coal-mines, lumber mills, etc., per day..	@ 20c.— 60.00	@ $1.00— 300.00	@ $3.00— 900.00
Engineer's wages, 300 days, per day.............	@$1.50 —450.00	@ 2.25— 675.00	@ 2.75— 825.00
Boy to switch, couple, etc.....	@ 60c.=180.00	@ 1.00— 300.00	@ 1.50— 450.00
Interest, 8 per cent., say..	250.00	250.00	250.00
Total	$970.00	$1,625.00	$2,625.00

There are a number of items which must be considered in a fair comparison of animals with locomotives, which vary too much with each individual case to be noted in the table given above.

A locomotive makes so much quicker time than animals, that fewer cars are required to carry a greater daily total of tonnage. This effects a reduction in original investment that may nearly amount to the cost of the locomotive, and also reduces materially the running expenses,

This reduction in the number of cars—the engine, with quick trips, replacing a number of teams making slow trips—reduces the number of turnouts needed. In one case one of our engines was mostly paid for by the sale of rails from extra track that was no longer of any use.

The keeping up of a path between the rails for animals to work on, the renewing of ties worn out by constant tramping over them, is a vexatious expense avoided by the use of a locomotive. This item often amounts to one man's continuous time, or $1 to $2 per day.

Even where a large sum is spent in keeping up a footway, the chance of accident and wear and tear of animals is greater, and the average useful life is less than that of a locomotive.

The relative economy increases rapidly with the length of the road. On a track of a quarter of a mile or less in length, the locomotive, although much preferable, would not have so much advantage as on a road half a mile long. While it is almost impracticable to haul with mules much over half a dozen miles, freight can be hauled ten miles by the locomotive cheaper than by mules two or three miles.

These incidental savings, which are not included in the table, will usually cover the additional cost if heavier rails are required, and also of any changes of grades, curves, mine headings, etc., as may be advisable for the most economical use of the locomotive.

We recommend that an engineer be also enough of a mechanic to do all light repairs and keep the locomotive in good order. With such a man, the item of repairs, unless the engine is over worked, should not average for, say 20 years, over $50 to $100 per year. The amount of fuel used is also considerably dependent on the engineer. We believe a liberal salary to a good, competent engineer the best policy. Our system of standard templets enables us to express duplicate parts on telegraphic orders. (See page 1.)

We believe that if parties who are doing hauling on tramways by animals will calculate for themselves the cost of operating, their own figures will show, more than ours, the advantages and economy of substituting light locomotives.

WEIGHTS OF LOGS AND LUMBER.

WEIGHT OF GREEN LOGS TO SCALE 1,000 FEET, BOARD MEASURE.

Yellow Pine (Southern)......... ....:. .8,000 to 10,000 lb.
Norway Pine (Michigan)........ ....7,000 to 8,000 lb.
White Pine (Michigan) { off of stump ....6,000 to 7,000 lb.
{ out of water.............................7,000 to 8,000 lb.
White Pine (Pennsylvania), bark off................................5,000 to 6,000 lb.
Hemlock (Pennsylvania), bark off.... ....6,000 to 7,000 lb.
Four acres of water are required to store 1,000,000 feet of logs.

WEIGHT OF 1,000 FEET OF LUMBER, BOARD MEASURE.

Yellow or Norway Pine........ ....Dry, 3,000 lb.; Green, 5,000 lb.
White Pine.............. ....Dry, 5,500 lb.; Green, 4,000 lb.

WEIGHT OF ONE CORD OF SEASONED WOOD, 128 CUBIC FEET PER CORD.

Hickory or Sugar Maple.................................... ....4,500 lb.
White Oak.................... .. ..3,850 lb.
Beech, Red Oak, or Black Oak... ....3,250 lb.
Poplar, Chestnut, or Elm..2,850 lb.
Pine (White or Norway)............................... ....2,000 lb.
Hemlock Bark, Dry (1 cord bark got from 1,500 feet logs)........ ....2,200 lb.

MEMORANDUM.—When wood is cut in 4 ft. lengths, a pile 4 ft. high and 8 ft. long contains one full cord of 128 cubic feet. Wood for locomotive fuel is cut in 2 feet lengths and a pile of 4 ft. high and 8 ft. long is reckoned as a locomotive cord. For our small locomotives wood should be cut about 18 inches long. The fuel reports of our wood-burning locomotives are given in locomotive cords of 64 cubic feet.

TO FIND THE SIZE OF RAIL NEEDED FOR A LOCOMOTIVE.

Multiply the number of tons (of 2,000 lb.) on one driving wheel by ten, and the result is the number of pounds per yard of the lightest rail advisable.

This rule is only approximate, and is subject to modification in practice. (NOTE.—If, as is often the case with four-wheel-connected locomotives, the weight on front and back driving wheels is not the same, the heavier weight must be taken.)

TO FIND THE NUMBER OF TONS OF RAIL PER MILE OF ROAD.

Multiply weight of rail per yard by 11, and divide by 7. This does not include sidings, and a ton is reckoned at 2,240 pounds.

EXAMPLE.—The number of tons of 28 pounds per yard rail required for one mile is 11 x 28=308 ; divided by 7=44 tons.

The number of tons of 2,000 pounds required per mile is very nearly 1¾ times the weight per yard.

EXAMPLE.—1¾ time gives 28 times 49 tons per mile required of 28 pounds rail.

Rails are regularly sold by the ton of 2,240 pounds.

TABLE OF TONS PER MILE REQUIRED OF RAILS OF FOLLOWING WEIGHTS PER YARD.

Weight per yard.	Tons of 2,240 lb. per mile.	Weight per yard.	Tons of 2,240 lb. per mile.
16 lb.	25 tons, 320 lb.	35 lb.	55 tons, 0 lb.
20 "	31 " 960 "	40 "	62 " 1,920 "
25 "	39 " 640 "	45 "	70 " 1,600 "
28 "	44 " 0 "	56 "	88 " 0 "
30 "	47 " 320 "	60 "	94 " 640 "

RAILROAD SPIKES, MADE BY DILWORTH, PORTER & CO., (LIMITED), PITTSBURGH, PENNA.

Size measured under head.	Average number, per keg of 200 lb.	Ties 2 ft. between centres, 4 spikes per tie, makes per mile.	Rail used, weight per yard.
5½ x 9/16	360	5,870 lb. = 29⅓ kegs.	45 to 70
5 x 9/16	400	5,170 " = 26 "	40 to 56
5 x ½	450	4,660 " = 23⅓ "	35 to 40
4½ x ½	530	3,960 " = 20 "	28 to 35
4 x ½	600	3,520 " = 17⅔ "	24 to 35
4½ x 7/16	680	3,110 " = 15½ "	20 to 30
4 x 7/16	720	2,940 " = 14¾ "	
3½ x 7/16	900	2,350 " = 11¾ "	16 to 25
4 x ⅜	1,000	2,090 " = 10½ "	
3½ x ⅜	1,190	1,780 " = 9 "	16 to 20
3 x ⅜	1,240	1,710 " = 8½ "	
2½ x ⅜	1,342	1,575 " = 7⅞ "	12 to 16

CROSS-TIES PER MILE.

Centre to centre.	Ties.
1½ feet.	3520
1¾ "	3017
2 "	2640
2¼ "	2348
2½ "	2113

SPLICE JOINTS PER MILE.

2 bars and 4 bolts and nuts to each joint.

Rails 20 feet long.	528 joints.
" 24 " "	440 "
" 26 " "	406 "
" 28 " "	378 "
" 30 " "	352 "

The length of rails as usually sold is 90 per cent. 30 feet long, and 10 per cent. 24 to 28 feet long, requiring 357 splice joints per mile.

Weights of splice joints vary according to their length, and also the size of bolts. The general shape of rails, as well as their weight per yard, also controls the weight of splice joints. Splice joints are sold both by the piece and by weight.

The average weight of splice joints (complete with 2 bars and 4 bolts and nuts) is as follows:

For rails of 16 to 20 lb. per yard, each joint weighs 5 to 6 lb.
" " 24 to 28 " " " " 6 to 8 "
" " 30 to 35 " " " " 10 to 12 "
" " 40 to 50 " " " " 12 to 16 "
" " 56 to 60 " " " " 18 to 24 "

WEIGHTS AND CAPACITIES OF CARS.

	NARROW GAUGE.		WIDE GAUGE.	
	Weight of car.	Weight of load.	Weight of car.	Weight of load.
8-wheel flat cars	6,500 lb. 8,500 lb.	20,000 lb. 30,000 lb.	16,000 to 18 000 lb. 17,000 to 19,000 lb. 18,000 to 20,000 lb. 19,000 to 21,000 lb. 20,000 to 23,000 lb. 23,000 to 25,000 lb.	24,000 lb. 28,000 lb. 30,000 lb. 40,000 lb. 50,000 lb. 60,000 lb.
8 wheel box cars...........	10,000 lb. 12,000 lb.	20,000 lb 30,000 lb.	19,000 to 20,000 lb. 19,000 to 21,000 lb. 20,000 to 24,000 lb. 26,000 to 28,000 lb. 23,000 to 30,000 lb.	24,000 lb. 30,000 lb. 40,000 lb. 50,000 lb. 60,000 lb.
4-wheel coal and ore cars...	4,000 lb. 6,000 lb.	10 000 lb. 12,000 lb.	7,000 lb. 9,000 lb.	16,000 lb. 20,000 lb.
8-wheel logging cars........ 4-wheel logging cars.	4,900 lb. 2,500 to 3,000 lb.	12.000 lb. (1,500 ft. of logs.) 10,000 lb. 12,000 lb.	5,600 lb. 5,000 lb. 6,000 lb.	20.000 lb. (2,500 ft of logs.) 16,000 lb. 20,000 lb.
Passenger coaches..........	20,000 to 22,000 lb.	46 to 64 passengers.	35,000 to 44,000 lb.	50 to 56 passengers.
Coaches for motor lines, suburban railroads, etc...	9,000 to 10,000 lb.	38 to 40 passengers seated; 75 to 100 passengers crowded.	10,000 to 14,000 lb.	40 to 50 passengers seated; 75 to 125 passengers crowded.
Open excursion coaches....	9,700 lb.	70 passengers.	9,700 lb. 18,000 lb.	70 passengers. 90 passengers.
One-horse car (16 ft. long) .. Two-horse car (23 ft. long).. 8-wheel street car...........	....	....	8,200 lb. 4,500 lb. 9,500 lb.	16 passengers. 22 passengers. 40 passengers.

The average weight of a passenger is 133 lbs., or 15 passengers per ton of 2,000 lb.

MISCELLANEOUS.

A bushel of bituminous coal weighs 76 pounds, and contains 2,688 cubic inches.

A bushel of coke weighs 40 pounds.

One acre of bituminous coal contains 1,600 tons of 2,240 pounds per foot of thickness of coal worked. Fifteen to 25 per cent. must be deducted for waste in mining.

A cubic yard of loose earth weighs 2,200 to 2,600 pounds.

A cubic yard of wet sand weighs 3,000 to 3,500 pounds.

A cubic yard of broken rock weighs 2,600 to 3,000 pounds.

Water weighs about $8\frac{1}{3}$ pounds per gallon, and one gallon contains 231 cubic inches.

One cubic foot contains almost exactly $7\frac{1}{2}$ gallons.

Cast iron weighs about 1 pound per 4 cubic inches.

Wrought iron weighs about one pound per $3\frac{1}{2}$ cubic inches.

The circumference of a circle is about $3\frac{1}{7}$ times its diameter.

One acre contains 43,560 square feet.

A square of $208\frac{7}{100}$ feet contains one acre $= 43,560$ square feet.

A square of $147\frac{81}{100}$ feet contains $\frac{1}{2}$ acre $= 21,780$ square feet.

A square of $104\frac{55}{100}$ feet contains $\frac{1}{4}$ acre $= 10,890$ square feet.

One square mile contains 640 acres.

To find the number of gallons in a circular tank multiply the diameter in feet by itself, then multiply by the depth in feet, then by 6, and from this sum deduct 2 per cent.

EXAMPLE.—A tank 14 feet diameter and 9 feet deep. 14 x 14 · 196 x 9 = 1764 x 6 = 10584 less 2% (= 210) = 10374 gallons. (This is very nearly exact.)

ESTIMATES OF COST OF ONE MILE OF RAILROAD TRACK.

Laid with steel rails weighing 16, 20, 25, 30, and 35 pounds per yard.

The following estimates are for the track ready for rolling stock, not including survey, right of way, buildings, tunnels, bridges, sidings, etc. They are intended merely to give a basis for more exact calculations, and will require modification to conform to variations in prices of material, freight charges, etc. The item of grading is very variable, and the lowest figures for this are for easy country, or where steep grades and curves are used to avoid expense in grading.

I.—Cost of one mile of track with 16 lb. steel rails.

	Rails at $32 per ton.	Rails at $37 per ton.	Rails at $42 per ton.
25$\frac{820}{2240}$ tons of 16 lb. steel rails	At $32 — $904.57	At $37 — $930.29	At $42 — $1,056.00
1,780 lb. of 3½ x ⅝ spikes	" 2½ c. — 44.50	" 2¾ c. — 48.95	" 3 c. — 53.40
357 splice joints..........	" 18 " — 64.26	" 20 " — 71.40	" 22 " — 78.54
2,640 cross ties..........	" 10 " — 264.00	" 15 " — 396.00	" 20 " — 528.00
Grading and laying track	400.00	— 600.00	— 900.00
Total per mile..........	$1,577.33	$2,046.64	$2,615.94

MEMO.—Each $1 per ton variation in the price of 16 lb. rails will make a difference of $25.14 per ton.

II.—Cost of one mile of track with 20 lb. steel rails.

	Rails at $30 per ton.	Rails at $35 per ton.	Rails at $40 per ton.
31$\frac{940}{2240}$ tons of 20 lb. steel rails	At $30 — $943.29	At $35 — $1,100.00	At $40 — $1,257.14
2,940 lb. of 4 x 7⁄16 spikes.	" 2¼ c. — 66.15	" 2⅜ c. — 69.83	" 2⅝ c. — 77.18
357 splice joints...... ...	" 20 " — 71.40	" 22 " — 78.54	" 24 " — 85.68
2,640 cross ties........ ..	" 10 " — 264.00	" 15 " — 396.00	" 20 " — 528.00
Grading and laying track	— 400.00	— 600.00	— 900.00
Total per mile..........	$1,744.84	$2,244.37	$2,848.00

MEMO.—Each $1 per ton variation in the price of 20 lb. rails will make a difference of $31.43 per mile.

III.—Cost of one mile of track with 25 lb. steel rails.

	Rails at $29 per ton.	Rails at $34 per ton.	Rails at $39 per ton.
$39\frac{640}{2240}$ tons of 25 lb. steel rails	At $29 $1,139.29	At $34 — $1,335.71	At $39 $1,532.14
3,520 lb. of 4 x ½ spikes..	" 2 c. — 70.40	" 2¼ c. = 79.20	" 2½ c. 88.00
357 splice joints.........	" 22 " = 78.54	" 24 " = 85.68	" 26 " 92.82
2,640 cross ties..........	" 10 " = 264.00	" 20 " = 528.00	" 30 " = 792.00
Grading and laying track	= 500.00	800.00	— 1,100.00
Total per mile...........	$2,052.23	$2,828.59	$3,604.96

MEMO.—Each $1 per ton variation in the price of 25 lb. rails will make a difference of $39.28 per mile.

IV.—Cost of one mile of track with 30 lb. steel rails.

	Rails at $28 per ton.	Rails at $33 per ton.	Rails at $38 per ton.
$47\frac{880}{2240}$ tons of 30 lb. steel rails.....	At $28 ... $1,320.00	At $33 $1,555.72	At $38 — $1,791.43
3,960 lb. of 4½ x ½ spikes.	" 2 c. 79.20	" 2¼ c. — 89.10	" 2½ c. — 99 00
357 splice joints..........	" 24 " = 85.68	" 26 " = 92.82	" 28 " = 99.96
2,640 cross ties........ ..	" 10 " = 264.00	" 20 " = 528.00	' 30 " = 792.00
Grading and laying track	500.00	= 900.00	= 1,200.00
Total per mile.....	$2,248.88	$3,165.64	$3,983.39

MEMO.—Each $1 per ton variation in the price of 30 lb. rails will make a difference of $47.14 per mile.

V.—Cost of one mile of track with 35 lb. steel rails.

	Rails at $27 per ton.	Rails at $32 per ton.	Rails at $37 per ton.
55 tons of 35 lb. steel rails	At $27 = $1,485.00	At $32 = $1,760.00	At $37 — $2,035.00
3,960 lb. of 4½ x ½ spikes	" 2 c. = 79.20	" 2¼ c. = 89.10	" 2½ c. = 99.00
357 splice joints....... ..	" 26 " = 92.82	" 28 " 99.96	" 30 " 107.10
2,640 cross ties......	" 10 ' = 264.00	" 25 " = 660.00	" 40 " = 1,064.00
Grading and laying track	= 600.00	= 1,000.00	= 1,200.00
Total per mile.........	$2,521.02	$3,609.06	$4,505.10

MEMO.—Each $1 per ton variation in the price of 35 lb. rails will make a difference of $55 per mile.

WORKING REPORTS.

The following record of work done by our locomotives is taken from reports furnished by their owners, excepting a few cases where our traveling agent has made tests. We take this opportunity of acknowledging our indebtedness to our customers who have taken so much trouble in furnishing us with this valuable and unique information.

These reports are not intended as a list of our locomotives in use, as a large proportion of our customers have never had a survey made and are unable to give the information. Many of these reports were made ten to fifteen years ago, and the conditions of service have been changed often meanwhile. In a few cases the same locomotive appears in the reports of different owners.

The average performance, and usually the best work done in regular service, is given, and this is generally considerably within the full capacity of the locomotive. The regular work is in some reports very much in excess of the estimated capacity, and in these cases there may be extra favorable conditions for overcoming grades by momentum, or the locomotive may be worked harder than usually advisable. In no case where a special test has been made with track and cars in good order, has any locomotive failed to come up to the estimated capacity.

These reports are not given as testimonials or recommendations, although we have an abundance of these, but our intention in presenting them is to give practical information, based on actual facts, instead of on theoretical calculations, as to the power, speed, daily mileage, and consumption of fuel and water of our locomotives ; and as to the grades and curves, the gauges of track, weights of rail and efficiency of different classes of roads on which light locomotives can be used advantageously.

We have placed these reports in tabular form, grouping together similar locomotives, arranged according to the sizes of cylinders and the steepness of the grades. By this arrangement a comparison can be made at a glance of the work done under various conditions. While these reports are necessarily unscientific, we know of no other record of locomotive performances that can be compared with them for practical use.

Memoranda of Work done by our Passenger Locomotives, from Reports Furnished by Owners.

Size of cylinders.	Page showing style.	Owner and Location, and Date of Report.	Gauge of track in inches.	Weight of rail in pounds per yard.	Length of road in miles.	Radius of sharpest curve in feet.	Grade in feet per mile.	Number of cars hauled at one time.	Weight of each car in pounds.	Load on each car in pounds.	Weight of train in tons of 2,000 lb.	REMARKS.
*7 x 12	10	South Florida R. R.... Sanford, Fla. (1881)	36 in.	16 lb.	22 mls.	......	20 ft.	5 cars	about 5,000 lb.	10,000 lb.	37 tons	Has hauled 7 cars, weighing 52 tons. Usual speed 20, and best 25 miles per hour with 3 loaded cars. 67 miles ¾ cord wood fuel, 3 tanks water per day, of 10 hours. Has made 140 miles per day of 14 hours. Monthly mileage, 1,725 miles.
7 x 12	37	Long Beach Development Co. Long Beach, Cal. (1888)	56½ in.	16 lb.	4 mls.	573 ft.	45 ft.	{ 4 cars / 1 car	2,000 lb. / 20,000 lb.	3,000 lb. / 30,000 lb.	} 35 tons	Frequently hauls 4 box cars—108 tons. Has hauled 4 small and 2 large coaches—58 tons. Regular speed 15, and fastest 25 miles per hour. 40 to 64 miles, 2 tanks of water. 200 lb. coal fuel per day of 14 hours.
8 x 12	18	Chino Valley R. R.... Chino, Cal. (1888)	42 in.	25 lb.	10 mls.	249 ft.	80 ft.	{ 1 car / 3 cars	16,000 lb. / 9,200 lb.	about 2,800 lb. empty	} 23 tons	Speed 17 miles per hour hauling 1 passenger car and 3 empty freight cars. 38 miles, burning 550 lb. of coal fuel and using 2½ tanks of water per day of 12 hours.
8 x 12	20	Mount Lookout Railway.... Chattanooga, Tenn. (1888)	36 in.	25 lb.	1¼ mls.	288 ft.	230 ft.	1 car	12,860 lb.	4,200 lb.	8½ tons	Has hauled 2 cars—17 tons. Regular speed 8, and fastest 20 miles per hour. 60 to 66 miles, 6 tanks of water, 1140 lb. soft coal per day of 23 hours.

Cyl.	No.	Name and Location	Gauge	Rail	Road			Cars	Weight per car	Total	Remarks
*8 x 16	10	Toledo & Maumee N. G. R. R., Toledo, Ohio. (1875)	36 in.	25 lb.	7½ mls.	250 ft.	52 ft.	{1 / 1}	passen-ger box car }	18 tons	1,500 lb. coal fuel, 90 miles per 12 hours; hauled 5 cars (53 tons), and ran 140 miles in excursion season. Best speed 30 miles per hour.
*8 x 16	10	Port Huron & N. W. R. R., Port Huron, Mich. (1880)	30 in.	30 lb.	46 mls.	1433 ft.	52 ft.	9 cars	11,000 lb. / 18,000 lb.	130 tons	3,000 lb. coal fuel, 184 miles per 12 hours; has run 296 miles in 14½ hours, and 5,200 miles per year. Usual speed 13 to 20, and best 30 miles per hour.
8 x 16	10	Peach Bottom R. W., E. Div., Oxford, Pa. (1875)	36 in.	30 lb.	28 mls.	300 ft.	105 ft.	5 cars	8,000 lb. / 12,000 lb.	50 tons	Grade 1 mile long; has hauled 6 freight and 1 passenger car. Ran 16 miles in 34 minutes.
8 x 16	36	Toledo & South Haven R. R., Lawton, Mich. (1881)	36 in.	30 lb.	13 mls.	600 ft.	130 ft.	{1 / 1}	passenger / freight car }	25 tons	Has hauled 6 cars—73 tons up 130 feet grade, and 8 empty and 2 loaded standard gauge cars on narrow gauge trucks, total weight 145 tons, up 60 feet grade. Usual speed with train of 2 cars 20, and best 30 miles per hour. 68 miles, 1 cord of wood fuel daily. This road carries on narrow gauge trucks loaded standard gauge cars a distance of 4 miles.
9 x 14	10	DeLand & St.Johns River R'y, DeLand, Fla. (1884)	36 in.	16 lb.	5 mls.	286 ft.	157 ft.	{1 / 3}	passenger / freight cars }	50 tons	30 to 50 miles daily, burning 1 cord of wood per day of 16 hours.
*9 x 10	8	Toledo, Delphos & I. R. R., Delphos, Ohio. (1877)	36 in.	30 lb.	16 mls.	......	26 ft.	8 cars	8,000 lb. / 16,000 lb.	96 tons	Has hauled 11 loaded cars. Regular speed 16 miles per hour, and usual mileage 64 to 96 miles per day.
*9 x 16	18	Coast Line R. R., Savannah, Ga. (1875)	60 in.	30 lb.	4 mls.	1150 ft.	42 ft.	0 passenger	cars	120 tons	½ cord pine wood, 3 tanks water, 56 miles per day.
*9 x 16	8	Havana, Rantoul & E. R. R., Rantoul, Ill. (1876)	36 in.	30 lb.	40 mls.	3700 ft.	52 ft.	10 cars	9,000 lb. / 14,500 lb.	188 tons	Has hauled 22 cars partly loaded, total weight about 200 tons. Mileage 70 to 140 miles per day. Usual speed 10 or 15 miles, and best speed 30 to 35 miles per hour.

* Change in owner, or location, or service, since report was made.

Memoranda of Work done by our Passenger Locomotives, from Reports Furnished by Owners.—Continued.

Size of cylinders.	Page showing style.	Owner and Location, and Date of Report.	Gauge of track in inches.	Weight of rail in pounds per yard.	Length of road in miles.	Radius of sharpest curve in feet.	Grade in feet per mile.	Number of cars hauled at one time.	Weight of each car in pounds.	Load on each car in pounds.	Weight of train in tons of 2,000 lb.	Remarks.
9 x 16	8	Clinton & Port Hudson R. R. Clinton, La. (1881)	56½ in.	40 lb.	22 mls.	337 ft.	57 ft.	7 cars	12,000 lb.	18,000 lb.	105 tons	44 to 48 miles, 1,000 lb. coal fuel daily. Principal business, carrying cotton. No trouble from sparks. Road built in 1883.
*9 x 16	8	East Line & Red River R. R. Jefferson, Texas. (1878)	36 in.	30 lb.	90 mls.	200 ft.	80 ft.	8 cars	9,000 lb.	16,000 lb.	100 tons	Has hauled 9 cars. Curve and grade come together. 100 mls. per day. Has run 30 miles per hour for three consecutive hours; greatest speed 45 miles per hour (36-inch drivers).
*10 x 16	19	Grand Haven R. R. Muskegon, Mich. (1879)	56½ in.			800 ft.	57 ft.	1 passenger	passenger	car	21 tons	Has hauled 5 cars = about 140 tons, at 30 miles per hour; has run 40 miles per hour. 112 miles, 1 cord wood fuel daily.
*10 x 16	8	Columbus, Wash. & Cin. R. R. Dayton, Ohio. (1880)	36 in.	30 lb.	20 mls.	.....	80 ft.	14 cars	10,000 lb.	16,000 lb.	182 tons	600 lb. coal per 40 miles run. Usual speed 10, and best 35 miles per hour. Has made 180 miles per day, and 3,590 miles per month.
*10 x 16	8	East Line & Red River R. R. Jefferson, Texas. (1878)	36 in.	30 lb.	90 mls.	200 ft.	80 ft.	11 cars	9,000 lb.	16,000 lb.	137 tons	Curve and grade come together; 100 miles per day.
10 x 16	19	Warrenton R. R. Warrenton, N. C. (1888)	56½ in.	30 lb.	3 mls.	240 ft.	92 ft.	{ 1 car, 1 car }	20,000 lb. 20,000 lb.	4,000 lb. 20,000 lb.	32 tons	Has hauled 6 loaded freight cars = 66 tons. Regular speed 20 and fastest 30 miles per hour. Grade ¾ mile long.

10 x 16	36	Ferrocarril de Hidalgo...... (1888) Mexico.	36 in.	30 lb.	90 mls.	300 ft.	95 ft.	3 cars	10,000 lb.	20,000 lb.	45 tons	50 and 150 miles per day, alternately.
*10 x 16	6	Cairo & St. Louis R. R...... St. Louis, Mo. (1877)	36 in.	40 lb.	147 mls.	573 ft.	95 ft.	4 passenger		cars	68 tons	Has hauled 5 cars—84 tons. Usual mileage 147 miles per day of 6½ hours, burning 2,000 lb. coal Regular 25, and best 40 miles per hour, except one run of 9 miles in 10 minutes. Ran 180 miles per day for a year or more, and ran 5,747 miles in one month.
10 x 16	6	Peach Bottom, R. W., M. Div. York, Pa. (1875)	36 in.	30 lb.	29 mls.	300 ft.	103 ft.	4 passenger		cars	50 tons	478 feet curve on grade; has hauled 6 cars—91 tons. 2,040 lb. coal, 1½ tanks, 144 miles,7½ hours. Has run 36 miles per hour (36-inch drivers). Ran 170 miles in one day, and 30,500 miles in one year.
10 x 16	8	Peach Bottom R. W., E. Div. (1875) Oxford, Pa.	36 in.	30 lb.	28 mls.	900 ft.	105 ft.	7 cars	8,000 lb.	{ 12,000 lb 18,000 lb	84 tons	93 to 129 miles. ½ ton coal and 2 tanks water per 12 hours.
10 x 16	8	Marietta & N'th Georgia R. R. Marietta, Ga. (1894)	36 in.	30 lb.	....	882 ft.	105 ft.	6 cars	12,000 lb.	18,000 lb.	90 tons	Grade 3 miles long; curve comes on grade. 142 miles, 3,400 lb. coal fuel, 5 tanks of water per day of 8 hours, running 7 days per week.
10 x 16	8	Ferrocarril de Cucuta...... U. S. of Colombia. (1884)	30⅝ in.	30 lb.	28¾ mls.	349 ft.	132 ft.	4 cars	10,097 lb.	13,508 lb.	47 tons	1⅚ cubic meters wood fuel and 2 tanks of water per trip of 44½ miles. Full power not tested.
10 x 16	8	Waynesburg & Wash. R. R.. Waynesburg, Pa. (1877)	36 in.	30 lb.	28½ mls.	185 ft.	137 ft.	5 cars	8,000 lb.	12,000 lb.	50 tons	Has hauled more. Curve is on grade. Has hauled 5 empty cars up straight grade of 240 feet per mile, and around 80 feet radius curve on temporary track.

* Change in owner, or location, or service since report was made.

Memoranda of Work done by our Passenger Locomotives, from Reports Furnished by Owners. – *Continued.*

Size of cylinders.	Page showing style.	Owner and Location, and Date of Report.	Gauge of track in inches.	Weight of rail in pounds per yard.	Length of road in miles.	Radius of sharpest curve in feet.	Grade in feet per mile.	Number of cars hauled at one time.	Weight of each car in pounds.	Load on each car in pounds.	Weight of train in tons of 2,000 lb.	REMARKS.
10 x 16	19	Sioux City & Highl'd P'k R'y, Sioux City, Ia. (1888)	58½ in.	30 lb.	5 mls.	288 ft.	158 ft.	3 cars	10,000 lb.	21,440 lb.	47 tons	Has hauled 4 passenger cars carrying 520 passengers. Also hauls freight. Runs 130 to 150 miles, burning 2,000 lbs. coal fuel, and using 3 tanks of water per day of 18 to 17 hours. Grade 1 mile long. No repairs in 12 months.
10 x 16	19	Winterville & Pleasant Hill R. R., Winterville, Ga. (1888)	57 in.	30 lb.	7 mls.	382 ft.	200 ft.	2 cars	15,000 lb.	40,000 lb.	55 tons	Grade 800 ft. long. Has hauled 4 cars—110 tons. Regular speed 12, and fastest 25 miles per hour. 84 to 168 miles, burning 3 cords engine wood per day of 12 hours when running 168 miles.
10 x 16	19	National City & Otay R. R., National City, Cal. (1888)	56½ in.	40 lb.	30 mls.	120 ft.	200 ft.	3 cars	10,000 lb.	12,000 lb.	38 tons	100 miles, 4 tanks of water, 2,000 lbs. coal fuel daily.
11 x 16	4	Suffolk & Carolina R. R., Suffolk, Va. (1888)	42 in.	35 lb.	40 mls.	716 ft.	66 ft.	25 cars	6,720 lb.	15,680 lb.	280 tons	Has hauled 35 cars—392 tons. Grade 1,200 feet long, with 4° curve. Usual speed 30, and best 40 miles per hour.
*11 x 16	6	Port Huron & N. W. R. R., Port Huron, Mich. (1881)	36 in.	30 lb.	118 mls.	1,433 ft.	53 ft.	16 cars	9,500 lb.	16,000 lb.	196 tons	Usual work is less. Has hauled 6 loaded cars 6 miles in 9 minutes up 53 feet grade.
		Same on return trip.....	36 in.	30 lb.	118 mls.	1,433 ft.	80 ft.	{ 8 1 1	9,500 lb. passenger car	16,000 lb baggage car	about 120 tons	140 miles, 3 tons coal daily. 3,050 miles average monthly mileage, exclusive of making up trains.
11 x 16	6	B. S. O. & B. R. R., Bedford, Ind. (16:8)	36 in.	35 lb.	41 mls.	382 ft.	85 ft.	7 cars	10,250 lb.	25,000 lb.	123 tons	Has hauled 8 cars. 12° reverse curve on steepest grade. 82 miles per 7½ hours. Frequent-

		Name, location (date)	Gauge	Rail	Length			Cars			Tons	Remarks
11 x 16	6	Chagrin Falls & Southern R'y, Chagrin Falls, O. (1888)	36 in.	30 lb.	5 mls.	....	95 ft.	6 cars	18,000 lb.	10,000 lb.	84 tons	ly hauls from quarries loads of 36,000 lbs. on a car, and hauled on 1 car a flagstone 22 feet 10 inches long, 11 feet wide and 2 feet thick, weighing 24 tons. Grade 1½ miles long. Has hauled 8 cars—152 tons. Regular speed 15, and fastest 20 miles per hour. 40 to 60 miles, 2 tanks water, 1,200 lbs. coal fuel daily.
*11 x 16	6	Hot Springs R. R., Hot Springs, Ark. (1878)	36 in.	35 lb.	24 mls.	318 ft.	150 ft.	{4 cars / 1 car}	9,000 lb. passenger	14,000 lb.	} 57 tons	Has hauled 8 cars—93 tons. Regular speed 18, and best 40 miles per hour over whole road. Usual mileage, 100 miles in 10 hours, best, 200 miles in 14 hours. 1½ cords wood and 2 tanks per 10 hours. Grade 1½ miles long.
12 x 18	18	Shell Beach R. R., New Orleans, La. (1885)	60 in.	42 lb.	30 mls.	955 ft.	10 ft.	10 cars	18,000 lb.	20,000 lb.	190 tons	Has hauled 19 cars—361 tons, 16 to 20 miles per hour : 120 to 180 miles, 6,000 lbs. coal fuel, 6 tanks of water, per day of 14 hours.
*12 x 18	6	Connotton Valley R. R., Canton, Ohio. (1880)	36 in.	35 lb.	22 mls.	288 ft.	110 ft.	3 cars	8,000 lb.	24,000 lb.	48 tons	1,300 lbs. coal fuel, 90 miles per day. Usual speed 15 to 30, and best 40 miles per hour.
12 x 18	19	Pasadena Railway, Pasadena, Cal. (1888)	36½ in.	35 lb.	7¼ mls.	573 ft.	116 ft.	2 cars	18,000 lb.	8,400 lb.	21 tons	58 miles, 2 tanks water, 140 gallons crude petroleum fuel. Has hauled 4 cars—89 tons. Road rises 500 feet in 7¼ miles.
12 x 16	4	Waynesburg & Wash. R. R., Waynesburg, Pa. (1881)	36 in.	30 lb.	28½ mls.	185 ft.	187 ft.	{1 / 1 / 4}	passenger baggage car 8,0 0 lb. 20,000 lb.	passenger car	} 63 tons	3,800 lbs. coal fuel, 112 to 140 miles per day. Curve is on grade.
12 x 16	16	St. Louis Cable & West. Ry. St. Louis, Mo. (1888)	36 in.	40 lb.	15 mls.	....	150 ft.	3 cars	16,000 lb.	16,000 lb. 8,000 lb.	42 tons	Has hauled 18 cars, carrying about 1,000 passengers, up grade of 105 feet per mile. Has run 228 miles in 31 hours, and made 25 miles per hour.
13 x 18	6	Hot Springs R. R., Hot Springs, Ark. (1881)	36 in.	35 lb.	22 mls.	531 ft.	158 ft.	{4 / 4}	freight passenger cars		98 tons	One ton coal fuel per 100 miles. 2 1-3 cords of wood, 90 to 150 miles per day. Usual speed 18, and best 35 miles per hour for regular runs.

* Change in owner, or location, or service, since report was made.

Memoranda of Work done by our Motors on Street and Suburban Railroads, from Reports furnished by Owners.

Size of cylinders.	Page showing style.	Owner and Location, and Date of Report.	Gauge of track in inches.	Length of track in miles.	Radius of sharpest curve in feet.	Grade in feet per mile.	Number of cars hauled at one time.	Weight of each car in pounds.	Load on each car in pounds.	Weight of train in tons of 2,000 lb.	REMARKS.
6 x 10	33	Wichita & Suburban R. R. Wichita, Kas. (1888)	56½ in.	....	46 ft.	53 ft.	8 cars	4-wheel street cars		12 tons	Use smoky soft coal.
6 x 10	33	Street Railway Co. Selma, Ala. (1888)	57 in.	5 mls.	90 ft.	about 53 ft.	1 car	4-wheel street car		3½ tons	Usual speed 6 miles, and fastest 25 miles per hour. 40 miles per day of 10 hours, burning 1-5 ton coke fuel, and using 1½ tanks of water. At 6 miles per hour, can stop in 6 feet. Costs $4.00 per day for engineer, coke and oil. Also hauls freight cars.
*6 x 10	33	Sou. California Motor R. R. San Bernadino, Cal. (1888)	36 in.	3½ mls.	....	80 ft.	1 car	5,000 lb.	9,800 lb.	7½ tons	96 to 103 miles, burning 600 lbs. of coal fuel per day of 14 hours.
7 x 12	33	New Orleans City R. R. New Orleans, La. (1878)	62½ in.	6 mls.	85 ft.	slight	6 cars	6,000 lb.	8,000 lb.	42 tons	Can haul more. Short grade of 1 foot per 100.
*7 x 12	32	Richmond City R. R. Richmond, Va. (1888)	57 in.	2¼ mls.	70 ft.	slight	2 cars	7,000 lb.	5,000 lb.	12 tons	Has hauled train weighing 18 tons. Usual speed 12, and fastest 22 miles per hour. 103 miles, burning 445 lbs. anthracite fuel, and using 3½ tanks of water per day of 13 hours. Has made 216 miles in 16 hours.
7 x 12	42	Wichita Rapid Transit. Wichita, Kansas. (1887)	42 in.	4 mls.	75 ft.	39 ft.	1 car	8,000 lb.	7,000 lb.	7½ tons	120 miles, burning ⅜ ton coal or coke, and using 7 to 9 tanks of water per day of 15 hours. On Sundays hauls 2 cars per trip, and once carried 400 people on 2 cars. Usual speed 12, and fastest 15 miles per hour. Have made 144 miles in 18 hours.

7 x 12	33	Johnstown & Stony Cr'k R.R., Johnstown, Pa. (1888)	56½ in.	3½ mls.	45 ft.	52 ft.	2 cars	10,000 lb.	6,000 lb.	16 tons	Has loop of 83 feet radius at end of track. 2,700 lbs. soft coal fuel, 4,000 gal. water, 180 miles per day of 20 hours. 44,625 miles in 8½ months.
7 x 12	33	Lakeside Street R. R., Canton, O. (1885)	48 in.	1⅜ mls.	60 ft.	73 ft.	1 car	4,000 lb.	7,000 lb.	5 tons	Frequently hauls 4 cars—24 tons, and has hauled 6 cars carrying 600 passengers—50 tons. Has run 1¼ mile in 8 minutes. 75 miles, burning 730 lbs. anthracite fuel per day of 13½ hours. Used in summer only, running 7 days per week without losing a trip. On busy days has done work that would require 70 horses.
*7 x 12	26	Miss. Val. & Ship Isl'd R. R., Vicksburg, Miss. (1881)	42½ in.	1½ mls.	286 ft.	90 ft.	1	passenger car		8 tons	Has hauled 5 freight cars—46 tons. 69 miles, burning 330 lbs. coal fuel daily. Used on branch line to the river landing.
7 x 12	32	Park Railway, Denver, Col. (1890)	42 in.	3½ mls.	40 ft.	90 ft.	1 car	30 ft.	car	8 tons	Round trips hourly.
7 x 12	32	Wichita & Suburban R. R., Wichita, Kansas. (1888)	56½ in.	....	75 ft.	158 ft.	1 car	4,000 lb.	4,000 lb.	4 tons	Grade about 500 feet long.
7 x 12	33	Lincoln Rapid Transit, Lincoln, Nev. (1887)	56½ in.	1⅜ mls.	....	211 ft.	2 cars	5,000 lb.	4,200 lb.	9 tons	Grade about 1,000 feet long. Burns about 800 lbs. anthracite fuel per 75 miles.
8 x 12	20	Tampa Street R. R., Tampa, Fla. (1888)	30 in.	2¾ mls.	143 ft.	8 ft.	3 cars	10,000 lb.	4,800 lb.	22 tons	Has hauled train of about 53 tons, full power not tested. 64 miles, burning ¾ cord of wood fuel, and using 8 tanks of water per day of 16 hours.
8 x 12	33	Seneca Falls & Wat'rloo R.R., Waterloo, N. Y. (1888)	56½ in.	3½ mls.	....	slight	1 car	10,000 lb.	7,000 lb.	8½ tons	126 miles per day.

* Change in owner, or location, or service, since report was made.

Size of cylinders.	Page showing style.	Owner and Location, and Date of Report.	Gauge of track in inches.	Length of track in miles.	Radius of sharpest curve in feet.	Grade in feet per mile.	Number of cars hauled at one time.	Weight of each car in pounds.	Load on each car in pounds.	Weight of train in tons of 2,000 lb.	REMARKS.
8 x 16	45	N. Orleans City & Lake R. R. New Orleans, La. (1878 and 1888)	62½ in.	7 mls.	90 ft.	slight	8 cars	19,000 lb.	8,400 lb.	110 tons	Usual speed 20 to 23 miles per hour; has run 3 miles in 5¾ minutes, with 2 cars. Has hauled 12 cars crowded and dead engine—213 tons, at 20 miles per hour. 100 miles, burning 1,200 lbs. anthracite fuel per day of 13 hours. Has made 161 miles in 23 hours.
9 x 14	42	St. Louis Cable & West. R.R. St. Louis, Mo. (1888)	36 in.	15 mls.	500 ft.	105 ft.	6 cars	8,000 lb.	15,000 lb.	69 tons	Usual speed 20, and fastest 30 miles per hour. 20 round trips of 1¾ miles and return per day of 10 hours.
9 x 14	44	East Birmingham Dummy Line. Birmingham, Ala. (1888)	56½ in.	7 mls.	95 ft.	210 ft.	2 cars	11,000 lb.	about 14,000 lb.	25 tons	Often hauls 3 cars crowded. Has hauled 6 cars. Have 11 curves of 95 feet radius. Usual mileage 160 miles per day of 16 to 18 hours. Has run 224 miles per day.
9 x 14	[33]	Butte Street Railroad. Butte, Montana. (1888)	56½ in.	2½ mls.	131 ft.	501 ft.	2 cars	3,600 lb.	5,000 lb.	7 tons	Has hauled 2 cars with 80 passengers—about 9 tons. Grade 400 feet long, with curve above and below. 90 miles, burning 1,200 lbs. coal fuel, and using 5 tanks of water per day of 18 hours.
9 x 14	33	Tacoma Street R. R. Tacoma, W. T. (1888)	42 in.	2 mls.	60 ft.	581 ft.	1 car	4,400 lb	6,000 lb.	5 tons	60 miles, ¾ ton coke or 1½ cord of 16-inch wood fuel, 5 tanks of water daily. 60 ft. curve on 9 per cent. grade. Has 7 grades, each 300 ft. long, varying from 9 to 11 per cent. in a distance of 2,500 feet.

9 x 14	44	Metropolitan Street R. R.... Atlanta, Ga. (1888)	50¼ in.	2 mls.	55 ft.	250 ft.	1 car	14,000 lb.	5,000 lb.	9½ tons	On busy days hauls 2 crowded cars—about 20 tons. Grades undulating, and steepest grade about 350 feet long, 55 feet radius curve is ¼-circle and on 182 feet per mile grade. Runs 17 hours per day, making round trip every 30 or 40 minutes. 7,517 miles in 3 months; 17¼ lbs. coke fuel per mile.
10 x 14	44	Dallas Rapid Transit Co..... Dallas, Tex. (1888)	56½ in.	....	200 ft.	79 ft.	4 cars	10,000 lb.	about 8,000 lb.	36 tons	Has hauled 4 loaded freight cars—127 tons. Grade ½ mile long, with 12-degree curve at summit.
10 x 14	44	Mil. & Whitefish Bay R'y..... Milwaukee, Wis. (1888)	56½ in.	4 mls.	95 ft.	158 ft.	3 cars	16,000 lb.	10,000 lb	89 tons	Has hauled 4 cars—52 tons. Usual speed 12, and fastest 15 miles per hour. Grade 800 feet long.
10 x 16	43	East End Railway......... Memphis, Tenn. (1888)	56½ in.	5½ mls.	127 ft.	185 ft.	2 cars	23,000 lb.	about 7,000 lb.	30 tons	Grade 700 feet long, road up and down hill, and rises 150 feet in 5½ miles. Has hauled 1 loaded freight car—45 tons. Usual mileage 16 trips—175 miles in 18 hours; has made 26 trips—286 miles in 18 hours. Does twice the work of other motors at half cost of repairs.
12 x 18	44	Columbus Railroad......... Columbus, Ga. (1888)	57 in.	8 mls.	126 ft.	175 ft.	7 cars	21,000 lb.	25,000 lb.	161 tons	Grade 500 feet long, and overcome by momentum. Speed 15 to 25 miles per hour. 72 to 98 miles, burning 1,500 lbs. coke fuel, and using 8 tanks of water per day of 12 hours.
12 x 18	43	Ensley Railway........ Birmingham, Ala. (1888)	56½ in.	7¾ mls.	57 ft.	211 ft.	5 cars	14,000 lb.	15,000 lb.	72½ tons	Grade 900 feet long. 124 and 170 miles alternately per day of 18 hours, burning 3,000 lbs. coke, and requiring 55 cents' worth of oil for 170 miles.
12 x 18	43	Mill. & Asylum Dummy Line Milledgeville, Ga. (1888)	56½ in.	4 mls.	70 ft.	264 ft.	2 cars	16,000 lb.	20,000 lb.	36 tons	Grade 600 feet long. Has hauled 4 cars—about 95 tons. 96 to 144 miles per day, burning 800 lbs. coal fuel.

* Change in owner, or location, or service, since report was made.

Memoranda of Work done by our Freight Locomotives, from Reports furnished by Owners.

Size of cylinders.	Page showing style.	Owner and Location, and Date of Report.	Gauge of track in inches.	Weight of rail in pounds per yard.	Length of track in miles.	Radius of sharpest curve in feet.	Grade in feet per mile.	Number of cars hauled at one time.	Weight of each car in pounds.	Load on each car in pounds.	Weight of train in tons of 2,000 lb.	REMARKS.
8 x 12	14	Astillero Plantation...... Mexico. (1888)	36 in.	18½ lb.	18 mls.	286 ft.	158 ft.	4 cars.	10,250 lb.	20,000 lb.	60 tons	Grade 4 miles long. Speed 9 to 13 miles per hour. 108 miles. 3 tanks water, 1¼ cords wood fuel per day of 12 hours.
9 x 14	19	Searcy & West Point R. R.... Searcy, Ark. (1888)	56½ in.	36 lb.	4½ mls.	358 ft.	60 ft.	{1 7}	12,500 lb. 22,500 lb.	35,000 lb.	208 tons	Has hauled train of 1 passenger coach, 8 loaded and 3 empty freight cars=270 tons. Grade ½ mile long. 45 to 63 miles daily, burning ¾ cord of wood fuel, and using 2 tanks of water. Speed 15 to 23 miles per hour.
9 x 14	36	Sandy River R. R...... Phillips, Maine. (1884)	24 in.	25 lb.	18 mls.	229 ft.	200 ft.	8 cars.	8,000 lb.	empty	32 tons	Sometimes hauls 9 cars. Cars carry 5 tons. Usual speed 15 miles per hour. 200-foot grade is ¼ mile long; loaded trains go up easier grades. 72 miles and ⅞ cord wood fuel daily.
9½ x 14	14	Martha's Vineyard R. R...... Edgartown, Mass. (1874)	36 in.	35 lb.	9 mls.	150 ft.	221½ ft.	3	passenger cars		30 tons	When busy runs 126 miles, burns 1,500 lbs. coal per day. Has run 30 miles per hour. Has hauled 4 cars with 420 passengers.

Cylinder	No.	Name, location (date)	Gauge	Rail	Length			Cars	Load	Load	Capacity	Remarks
*9½ x11	14	Pine R. Val. & Stevens Pt. R.R. Richland Centre, Wis. (1876)	36 in.	......	16 mls.	318 ft.	53 ft.	5 cars	10,000 lb.	10,000 lb.	50 tons	For some years used maple-wood rail. Has hauled 7 cars. Regular mileage 64 miles per 8 hours, burning 2¾ cords wood.
*9½ x14	14	Chester & Lenoir N. G. R. R. Yorkville, S. C. (1875)	36 in.	35 lb.	20 mls.	820 ft.	65 ft.	{2 4}	passenger } freight cars		55 tons	Has hauled 8 cars—73 tons. Grade 1 mile long. ⅓ cord wood, 2 tanks water per day. 40 miles per 4 hours.
9¾ x14	14	Peach Bottom R. W., M. Div. York, Pa. (1875)	36 in.	80 lb.	28 mls.	900 ft.	105 ft.	3 cars	7,500 lb.	18,000 lb.	88 tons	Grade 6 miles long with 500-foot curve. Has hauled 70 tons. 60 miles, 1,140 lbs. coal, 1 tank water per day. Has made 112 miles.
*9½ x14	14	Santa Cruz & Felton R. R. Santa Cruz, Cal. (1875)	36 in.	...	8 mls.	191 ft.	187 ft.	5	passenger cars		60 tons	Special trial. Have only 3½ miles straight track.
9½ x14	80	Arizona Copper Co. (Ltd.) Clifton, Arizona (1884)	20 in.	24 lb.	6¾ mls.	144 ft.	211 ft.	16 cars	2,850 lb.	empty	24 tons	Grade 3,000 feet long. Cars carry 5,000 lbs. each coming down grade. 44 to 55 miles, 1¼ cords cedar fuel, 8 tanks water daily. Has tank on boiler.
9½ x14	14	Rising Fawn Iron Works Dade Co., Ga. (1875)	30 in.	30 lb.	39¾ mls.	75 ft.	316 ft.	10 cars	4,000 lb.	empty	20 tons	30 miles, 4 tanks water, 2,000 lbs. coal per day.
*9 16	21	Cincinnati & Green Riv. R.R. King's Mountain, Ky. (1881)	36 in.	......	......	102 ft.	182 ft.	5 cars	6,500 lb.	16,000 lb	61 tons	Has hauled 8 cars—90 tons. Speed 15 to 25 miles per hour. 50 to 98 miles per day.
10 x 16	10	Clarksb'g, W. & G.R.R. & T.Co. Weston, W. Va. (1880)	36 in.	80 lb.	25¾ ml.	716 ft.	105 ft.	6 cars	9,800 lb.	15,000 lb.	73 tons	108 miles, 2,800 lbs. coal per day of 8 hours. Usual speed 15, and best 20 miles per hour. Has hauled 8 cars—81 tons. Grade 2 miles long. Curve of about 150 feet radius on siding.

* Change in owner, or location, or service, since last report was made.

Memoranda of Work done by our Freight Locomotives, from Reports furnished by Owners.—*Continued.*

Size of cylinders.	Page showing style.	Owner and Location, and Date of Report.	Gauge of track in inches.	Weight of rail in pounds per yard.	Length of road in miles.	Radius of sharpest curve in feet.	Grade in feet per mile.	Number of cars hauled at one time.	Weight of each car in pounds.	Load on each car in pounds.	Weight of train in tons of 2,000 lb.	REMARKS.
12 x 16	16	Albemarle & Pantego R. R., Norfolk, Va. (1888)	56½ in.	50 lb.	30 mls.	578 ft.	25 ft.	10 cars	17,000 lb to 28,000 lb	35,000 lb to 45,000 lb	300 tons	Has hauled 20 cars=500 tons on nearly level track. Speed 20 to 30 miles per hour.
12 x 18	12	Campbell's Creek Coal Co., Malden, W. Va. (1880)	56½ in.	40 lb.	3 mls.	100 ft.	22 ft.	60 cars	4,000 lb.	3,000 lb.	210 tons	Has hauled 100 cars=about 350 tons. 30 miles, 2,000 lbs. coal per day.
12 x 16	16	Poronai R. W. of Hokkaido, Yesso, Japan. (1881)	42 in.	30 lb.	3 mls.	320 ft.	53 ft.	19 cars	8,000 lb.	12,000 lb.	228 tons	Full power not tested. Curve is in tunnel.
12 x 16	16	Boston & Colorado Smelt. Co., Denver, Col. (1880)	36 in.	30 lb.	25 mls.	191 ft.	75 ft.	12 cars	9,000 lb.	19,000 lb.	168 tons	Connects with Colorado Central R. R. Grade 1½ miles long. Also does switching.
12 x 16	16	Houston, E. & W. Texas R.R., Houston, Tex. (1881)	36 in.	35 lb.	100 mls.	1910 ft.	90 ft.	26 cars	about 9,000 lb.	22,500 lb.	410 tons	Has hauled 30 cars weighing about 435 tons, 12 to 20 miles per hour. 100 miles, 3 cords wood fuel, 3 tanks water per day of 10 hours.
12 x 16	16	B. S. O. & B. R. R......, Bedford, Ind. (1878)	36 in.	35 lb.	42 mls.	383 ft.	85 ft.	12 cars	10,250 lb.	16,000 lb.	159 tons	Has frequently hauled 16 loaded cars=210 tons. Ran 184 miles with 2.310 lbs. coal. Annual mileage 24,666 miles.

	No.	Owner, or location, or service	Gauge	Rail	Road			Cars			Weight	Remarks
12 x 18	16	Arizona & New Mexico R'y.. Lordsburg, N. M. (1884)	36 in.	35 lb.	75 mls.	200 ft.	105 ft.	9 cars	12,000 lb.	14,000 lb.	117 tons	Usual speed per hour 15 miles on easy grades, and 10 miles on heavy grades; best speed 35 miles, 2,000 lbs. coal fuel and 2 tanks of water each trip of 71 miles. Grade 5 miles long; curve comes on grade.
12 x 18	19	De Bardeleben Coal & Iron Co. Bessemer, Ala. (1888)	50⅛ in.	56 lb.	4 mls.	280 ft.	100 ft.	0 cars	23,000 lb.	40,000 lb.	189 tons	
12 x 18	16	Black Hills R. R...... Lead City, Dakota. (1884)	36 in.	30 lb.	17 mls.	160 ft.	116 ft.	10 cars	7,000 lb.	16,000 lb.	115 tons	Has hauled 12 cars weighing 188 tons. Grade ¾ mile long. Usual speed 12 miles per hour; has run 20 miles per hour. 85 to 108 miles, 3 cords wood fuel, and about 3 tanks water daily. Curve one half circle long. This locomotive hauls cord-wood to the top of a chute 1,500 feet long, curved at the bottom to pile wood 100 feet high.
12 x 18	10	The Iron & Steel Works Ass'n of Virginia (Ltd.) Goshen, Va. (1884)	36 in.	35 lb.	12 mls.	410 ft.	147 ft.	10 cars	7,500 lb.	empty	37½ tons	Has hauled 14 cars weighing 52¼ tons. Grade 1,500 feet long, with curve of 500 feet radius. 72 to 120 miles, 3,000 lbs. coal, 2 tanks water daily. Usual speed 15, and best 30 miles per hour.
		Same on trestle work at stock house, track and grade 700 feet long.	.....	.....	.....	.....	204 ft.	3 cars	7,500 lb.	22,400 lb.	45 tons	
12 x 18	16	Black Hills R. R...... Lead City, Dakota. (1884)	36 in.	30 lb.	17 mls.	160 ft.	150 ft.	5 cars	7,000 lb.	16,000 lb.	57 tons	Curve of 181 feet radius on grade of 120 feet per mile. Many of the 160 feet curves occupy ⅜ circle. Has 110 feet radius curve on a Y. Empty trains go up 211 feet grade.

* Change in owner, or location, or service, since report was made.

Memoranda of Work done by our Freight Locomotives, from Reports furnished by Owners.—Continued.

Size of cylinders.	Page showing style.	Owner and Location, and Date of Report.	Gauge of track in inches.	Weight of rail in pounds per yard.	Length of road in miles.	Radius of sharpest curve in feet.	Grade in feet per mile.	Number of cars hauled at one time.	Weight of each car in pounds.	Load on each car in pounds.	Weight of train in tons of 2,000 lbs.	Remarks.
*12 x 16	12	BinghamCan.&Ca.FloydR.R. Salt Lake, Utah. (1875)	36 in.	30 lb.	15 mls.	......	186 ft.	10 cars	7,000 lb.	empty	35 tons	Has hauled 75 tons. · Grade 1 mile long; has short grades steeper.
*12 x 16	16	Colorado Central R. R..... Golden, Col. (1873)	36 in.	35 lb.	25 mls.	143 ft.	191 ft.	4 cars	10,500 lb.	16,000 lb.	53 tons	Has hauled 5 cars. Has 272 feet grade 125 feet long. 4,000 lbs. brown coal and 1 cord wood, 5 tanks, 65 miles per 6½ hours.
12 x 16	16	Nashville & Tuscaloosa R. R. Dickson, Tenn. (1881)	36 in.	40 lb.	25 mls.	143 ft.	211 ft.	6 cars	7,800 lb.	16,000 lb.	71 tons	Has 151 feet curve on 158 feet grade. The 211 feet grade only 200 feet long. Has 185 feet grade 1,000 feet long.
12 x 18	21	Bodie & Benton R. W. & Commercial Co. Bodie, California. (1882)	36 in.	35 lb.	38 mls.	286 ft.	211 ft.	3 cars	8,825 lb.	20,000 lb.	43 tons	12 miles per hour. 108 to 144 miles, 1½ cords wood fuel, 4 tanks water per day of 12 hours. Have 7 miles of grade 180 to 211 feet per mile, with 3 switch-backs. Has hauled 4 cars=50 tons.
12 x 16	16	Ferrocarril de Zaza......... Zaza, Cuba. (1879)	36 in.	30 lb.	25 mls.	640 ft.	224 ft.	5 cars	7,000 lb.	16,000 lb.	57 tons	Has hauled 6 cars=69 tons. 50 to 100 miles, 1½ cords wood fuel per day.

Cylinders	No.	Name and location (date)	Gauge	Rail weight	Length of road	Radius	Grade	No. cars	Freight passenger cars	Empty cars	Load	Remarks
*12 x 16	12	Wasaatch & Jordan R. R... Salt Lake, Utah. (1875)	36 in.	30 lb.	11 mls.	......	287 ft.	7 / 2			40 tons	
*12 x 16	12	American Fork R. R........ Salt Lake, Utah. (1874)	36 in.	30 lb.	15 mls.	240 ft.	300 ft.	9 cars	4,500 lb.	empty	15 tons	Hauled 47 tons on special trial. Road not now in use.
12 x 18	21	Callie Furnace .... Williamsons, Va. (1881)	56½ in.	35 lb.	5 mls.	109 ft.	300 ft.	1 car	20,000 lb.	28,000 lb.	24 tons	Has hauled 11 empty 4-wheel cars—41 tons. 1,080 lbs. coal, 40 to 60 miles daily.
*13 x 18	10	Texas & St. Louis R. W ...... Tyler, Texas. (1881)	36 in.	......	.....	......	70 ft.	21 cars	10,000 lb.	21,000 lb.	325 tons	Grade ¾ mile long.
13 x 18	16	Danville & New River R. R.. Danville, Va. (1884)	35 in.	35 lb.	61 mls.	388 ft.	150 ft.	8 cars	15,000 lb.	20,000 lb.	140 tons	Grade 2,400 feet long, 478 feet radius curve on 150 feet grade. 123 miles, 3 cords wood fuel, 2 tanks of water per day of 10 hours. Has hauled 9 cars weighing 157 tons. Usual speed 15 miles; no test made for speed. Also hauls mixed trains of 2 passenger and 3 to 5 freight cars.
14 x 20	16	Prescott & Arizona Cent.R.R. Prescott, Ariz. (1888)	56½ in.	40 lb.	74 mls.	......	105 ft.	5 cars	20,000 lb.	98,000 lb.	140 tons	Has hauled 7 cars—210 tons. Usual speed 18, and fastest 20 miles per hour. 143 miles, 2 tanks water, 4½ tons coal fuel per 9 hours.
14 x 20	18	Home Avenue R. R.......... Dayton, O. (1886)	56½ in.	35 lb.	8 mls.	200 ft.	186 ft.	4	loaded	cars	104 tons	Special trial.
14 x 20	21	Lime Rock R. R............. Rockland, Me. (1889)	56½ in.	50 lb.	8 mls.	301 ft.	227 ft.	12	loaded	cars	84 tons	Pushed train up grade with ease.

* Change in owner, or location, or service, since report was made.

Memoranda of Work done by our Six-Wheel-Connected Tank Locomotives, from Reports furnished by Owners.

Size of cylinders.	Page showing style.	Owner and Location, and Date of Report.	Gauge of track in inches.	Weight of rail in pounds per yard.	Length of road in miles.	Radius of sharpest curve in feet.	Grade in feet per mile.	Number of cars hauled at one time.	Weight of each car in pounds.	Load on each car in pounds.	Weight of train in tons of 2,000 lb.	REMARKS.
9 x 14	22	Edgemoor Iron Co., Wilmington, Del. (1884)	36 in.	30 lb.	.....	50 ft.	about 53 ft.	2 cars	20,800 lb.	40,000 lb.	60 tons	180 feet radius curve on grade. 360 lbs. coal fuel, 1 tank of water daily. Used for yard work.
*9 x 14	22	Keller & Reilly (contractors), Lancaster, Pa. (1881)	36 in.	old 56 lb.	2 mls.	143½ ft.	150 ft.	14 cars	1,500 lb.	4,000 lb.	38 tons	40 miles, 800 lbs. coal fuel daily.
*9 x 14	22	Dade Coal Co., Cole City, Ga. (1874)	36 in.	25 lb.	2½ mls.	164 ft.	158 ft.	35 cars	1,500 lb.	empty	26 tons	Has hauled 65 cars=49 tons. Hardest work holding back loaded trains. 50 to 65 miles daily.
*8¼ x 16	22	Kittle & Co. (contractors), Keokuk, Ia. (1873)	36 in.	28 lb.	2 mls.	.....	176 ft.	20 cars	1,600 lb.	3,600 lb.	52 tons	Short 264 and 290 feet grades. Could haul 30 cars; 15 to 20 bushels Iowa coal fuel, 40 miles daily.
*9 x 14	22	Lawrence Ore Co. (Ltd.) Wampum, Pa. (1883)	36 in.	.....	.....	.....	237 ft.	10 cars	4,000 lb.	empty	20 tons	Has hauled 13 cars weighing about 26 tons, starting on grade; grade about ⅛ mile long. Also hauls 6 empty cars weighing 12 tons up short grade of 335 feet per mile.

*8¼ x 16	22	Chas. McFadden (contractor) Downingtown, Pa. (1878)	30 in.	......	......		about 265 ft.	0 cars	1,600 lb.	empty	15 tons	Contractor's road, rough track, and constantly shifted.
9 x 14	22	M. S. Marquis (quarry) New Castle, Pa. (1890)	36 in.	40 lb.	1⅛ mls.	45 ft.	395 ft.	16 cars	1,000 lb.	empty	18 tons	Has hauled 22 cars—20 tons. Curve of over ⅜ circle, varying from 45 to 80 feet radius on a grade about 300 feet long, varying from 300 to 400 feet per mile. Road rises 300 feet in 1⅛ miles. 27 to 36 miles daily. Can get out 800 tons limestone daily.
		Same on different track......	86 in.	40 lb.	......	85½ ft.	204 ft.	4 cars	1,900 lb.	6,000 lb.	16 tons	Sometimes hauls 5 loaded cars of fire-clay—20 tons.
9 x 16	22	West Virginia Central & Pittsburgh R. R. Piedmont, W. Va. (1864)	41 in.	30 lb.	1¼ mls.	358 ft.	95 ft.	26 cars	2,000 lb.	empty	26 tons	Has hauled 34 cars, speed 9 to 12 miles per hour, 45 to 65 miles per day of 11 to 13 hours. 2,000 lbs. coal fuel, 4 tanks water daily. This locomotive is used on track from mouth of mine to top of slope, and gets out 950 tons coal daily when busy.
9½ x 14	22	Knoxville & New River R. R. Robbins, Tenn. (1884)	36 in.	25 lb.	18 mls.	320 ft.	210 ft.	3 cars	6,100 lb.	11,000 lb.	26 tons	No special test made. Road-bed under construction and not ballasted.
9½ x 14	22	Plantation of Sr. Agustin de Ariosa, Cuba. (1881)	36 in.	30 lb.	3 mls.	......	about 50 ft.	3 cars	6,000 lb.	12,500 lb.	27 tons	Often hauls 6 cars—55 tons. About 90 miles daily mileage.

* Change in owner, or location, or service, since report was made.

Memoranda of Work done by our Six-Wheel-Connected Tank Locomotives, from Reports furnished by Owners.—*Continued.* (104)

Size of cylinders.	Page showing style.	Owner and Location, and Date of Report.	Gauge of track in inches.	Weight of rail in pounds per yard.	Length of road in miles.	Radius of sharpest curve in feet.	Grade in feet per mile.	Number of cars hauled at one time.	Weight of each car in pounds.	Load on each car in pounds.	Weight of train in tons of 2,000 lb.	REMARKS.
*9½x14	22	Sloss Furnace Co., Birmingham, Ala. (1884)	36 in.	32 lb.	1¼ mls.	....	158 ft.	12 cars	4,000 lb.	7,000 lb.	66 tons	Can handle 16 to 18 cars if necessary.
*9½x14	22	College Hill R. R., Cincinnati, Ohio (1878)	36 in.	35 lb.	6½ mls.	235 ft.	249 ft.	1 car	19,000 lb.	6,000 lb.	12 tons	Has hauled 3 cars=28 tons. 84 to 108 miles, 1,500 lbs. coal fuel, 3 tanks water per day.
9½ x 14	22	Hawk's Nest Coal Co., Ansted, W. Va. (1881)	30 in.	35 lb.	4 mls.	100 ft.	348 ft.	16 cars	2,000 lb.	empty	16 tons	Has hauled 20 cars =20 tons. 72 to 80 miles, 2,400 lbs. coal fuel, and 8 tanks of water daily. More than one half of the road is made up of 22° curves.
9½ x 14	22	Potomac Coal Co., Barton, Md. (1884)	48 in.	30 lb.	1½ mls.	....	370 ft.	18 cars	1,800 lb.	empty	16 tons	Gets out 900 tons coal daily.
*10 x 16	23	Rorer Iron Co., Roanoke, Va. (1884)	36 in.	30 lb.	5½ mls.	193 ft.	158 ft.	6 cars	8,500 lb.	10,000 lb.	55 tons	Has hauled 7 cars weighing 60 tons. Have 317 feet grade 900 feet long for empty cars to come up. Regular speed 15, and best 25 miles per hour. 55 to 60 miles, 1,000 lbs. coal fuel, 5 tanks of water daily.
10 x 16	23	Virginia Steel Co., Roanoke, Va. (1889)	36 in.	30 lb.	6 mls.	125 ft.	422 ft.	8 cars	3,000 lb.	empty	12 tons	Grade short; curve and grade coming together. 48 to 60 miles, 1,500 lbs. coal fuel, daily. Also hauls loaded trains of 60 tons up grades of 237 ft. per mile, 400 ft. long, with reversed curves of 195 ft. radius on grade.

*10 x 10	23	College Hill R. R...... Cincinnati, Ohio. (1881)	36 in.	35 lb. 6½ mls.	235 ft.	240 ft.	3 cars	14,000 lb.	14 000 lb	42 tons	One round trip each hour, 1,000 lbs. coal fuel, 4 tanks of water daily.
10 x 16	23	Michigan Slate Co.... Avon, Mich. (1884)	36 in.	35 lb. 5½ mls.	358 ft.	353 ft.	0 cars	2,000 lb.	6,000 lb.	24 tons	Usual work, empty cars up 353 feet grade, except when supplies, etc., are carried. Cars coming down carry 10,000 lbs. slate each. Usual speed 12 miles. Ran 30 miles per hour with 2 loaded cars. 66 miles, 1½ cords wood fuel, 2 tanks of water daily.
12 x 18	23	Home Avenue R. R........ Dayton, Ohio. (1881)	50½ in.	35 lb. 3 mls.	200 ft.	180 ft.	4 cars	7,000 lb.	15,000 lb.	44 tons	Has hauled 60 tons. Usual speed 15, and best 30 miles per hour. 66 to 138 miles, 1,500 lbs. coal fuel daily.
12 x 16	23	Profile & Franconia Notch Railroad. White Mountains, N. H. (1881)	36 in.	40 lb. 3½ mls.	......	230 ft.	3 cars	20,000 lb.	6,000 lb.	39 tons	56 miles daily. This branch road is used chiefly to take the place of a stage line for summer travel.
*12 x 18	23	Poplar Mountain Coal Co... Cumberland City, Ky. (1879)	42 in.	40 lb. 7½ mls.	220 ft.	300 ft.	12 cars	5,700 lb.	empty	34 tons	Can haul 16 cars. Usual mileage 45, best 75 miles per day. 600 lbs. coal and 1 tank water per 15-mile trip.

* Change in owner, or location, or service, since report made.

Memoranda of Work done by our Shifting Locomotives, from Reports furnished by Owners.

Size of cylinders.	Page showing style.	Owner and Location, and Date of Report.	Gauge of track in inches.	Radius of sharpest curve in feet.	Grade in feet per mile.	Number of cars hauled at one time.	Weight of each car in pounds.	Load on each car in pounds.	Weight of train in tons of 2,000 lbs.	REMARKS.
*9 x 16	24	Duquesne Coal Co., Pittsburgh, Pa. (1875)	56½ in.	120 ft.	110 ft.	5 cars	15,200 lb.	empty	38 tons	Grade 300 feet long. Has hauled 12 cars. 270 lbs. coal fuel, 3 tanks water per day.
*9 x 16	24	Jackson Iron Co., Michigan. (1873)	56½ in.	152 ft.	154 ft.	6 cars	6,000 lb.	12,000 lb.	54 tons	Curve on grade. Former load on wood rail, 3 cars.
9 x 16	24	New Castle R.R. & Mining Co., New Castle, Pa. (1889)	42 in.	....	271 ft.	18 cars	2,200 lb.	empty	20 tons	Locomotive built in 1867, and the first locomotive built by Smith & Porter.
10 x 16	24	Belmont Nail Works, Wheeling, W. Va. (1876)	56½ in.	70 ft.	slight	10 cars	18,000 lb.	20,000 lb.	190 tons	Track only 200 yards long.
10 x 16	24	Pittsburgh Steel Casting Co. (1888)	56½ in.	....	about 26 ft.	10 cars	20,000 lb.	30,000 lb	220 tons	This locomotive is also used for charging a cupola, hauling a specially-constructed car with weigh scales, metal and fuel, weighing about 18 tons, up an incline 175 feet long with a grade of 6 per 100. Hot ingots are also hauled through the works.
*10 x 16	24	Fort Dodge Coal Co., Fort Dodge, Ia. (1978)	56½ in.	....	30 ft.	12 cars	16,000 lb.	20,000 lb.	216 tons	Hauled 23 cars=414 tons on 30 feet grade; also ran 100 miles in 7 hours on special trial. 1,000 lbs. coal fuel, 3 tanks water, 24 to 36 miles per day.
		Same on different track.	56½ in.	....	80 ft.	6 cars	16,000 lb.	20,000 lb	108 tons	
*10 x 16	24	Atkins Bros., Pottsville, Pa. (1876)	56½ in.	150 ft.	132 ft.	6 cars	6,000 lb.	10,000 lb.	48 tons	Curves on grades. Best work, 64 tons up 132 feet, and 32 tons up 185 feet grade. 750 lbs. anthracite fuel, 2 tanks water per day.
		Same on different track.	56½ in.	150 ft.	185 ft.	3 cars	6,000 lb.	10,000 lb.	24 tons	

		Owner, or location				Cars	Weight	Empty	Tons	Remarks
10 x 16	24	Munhall Bros ..... Pittsburgh, Pa. (1881)	43½ in.	....	139 ft.	50 cars	about 1,200 lb.	empty	80 tons	Round trip of 4½ miles in 35 minutes. Has hauled 80 empty cars 2¾ miles up grade in 13 minutes. 159 feet per mile grade is 1 mile long. 1,400 lbs. coal fuel daily. Full power not tested.
10 x 16	24	Bellaire Nail Works..... Bellaire, O. (1888)	36 in.	60 ft.	211 ft.	6 cars	6,000 lb.	8,800 lb.	20 tons	Grade 200 feet long. Taking a run at grade has hauled 12 cars—58 tons. Also hauls fluid metal from furnace to converter.
10 x 16	24	Sobo Furnace..... Pittsburgh, Pa. (1881)	56½ in.	123 ft.	231 ft.	3 cars	20,000 lb.	40,000 lb.	90 tons	Grade 120 feet long on reversed 123 feet curves.
10 x 16	24	Singer Manufacturing Co..... Elizabeth, N. J. (1884)	56½ in.	110 ft.	288 ft.	5 cars	6,400 lb.	12,000 lb.	46 tons	Has hauled 7 cars weighing 65 tons. 1,000 lbs. coal fuel, 4 tanks water daily.
10 x 16	24	Crozer Steel & Iron Co..... Roanoke, Va. (1884)	36 in.	191 ft.	322 ft.	8 cars	4,000 lb.	empty	16 tons	322 feet per mile grade is 300 feet long. Road 3 miles long, average grade 238 feet per mile. Has hauled 10 cars weighing 20 tons. Usual speed 15, and best 30 miles per hour. 42 to 48 miles, 900 lbs. coal fuel, 4 tanks of water per day of 10 hours.
*11 x 10	24	Kittaning Coal Co..... C. B. Finley, M'k'r. (1884) Houtzdale, Pa. (Memo. The apparatus is patented by Mr. Finley, to whom all are referred for further details.)	36 in.	100 ft.	50 ft.	35 cars	1,900 lb.	8,000 lb.	98 tons	Has hauled 50 cars weighing 122 tons. This locomotive does this work not by direct traction, but working on friction rollers which operate a wire rope. After the locomotive has served the purpose of a stationary engine and pulled the cars up out of the mine, it couples onto the train and hauls it about one mile. Two mines are thus worked. 38 to 40 miles, 800 lbs. coal fuel, 4 tanks of water daily.
12 x 18	24	Haskell & Barker Car Co..... Michigan City, Ind. (1888)	56½ in.	120 ft.	60 ft.	8 cars	28,000 lb.	40,000 lb.	252 tons	Can shift 10 or 12 cars easily. About 1 tank of water and 600 lbs. coal fuel daily. Shifts 50 to 60 cars daily.
*12 x 18	24	Ross & Sanford, contractors, Potomac Flats, (1884) Washington, D.C.	56½ in.	450 ft.	61 ft.	10 cars	10,000 lb.	20,000 lb.	150 tons	About 1½ tons coal fuel daily.
*12 x 18	24	Lochiel Iron Co..... Harrisburg, Pa. (1876)	56½ in.	....	182 ft.	8 cars	18,000 lb.	27,000 lb.	190 tons	Has hauled 500 tons on siding. Grade slight and not measured.

* Change in owner, or location, or service, since report was made.

(107)

Memoranda of Work done by our Shifting Locomotives, from Reports furnished by Owners.—Continued.

Size of cylinders.	Page showing style.	Owner and Location, and Date of Report.	Gauge of track in inches.	Radius of sharpest curve in feet.	Grade in feet per mile.	Number of cars hauled at one time.	Weight of each car in pounds.	Load on each car in pounds.	Weight of train in tons of 2,000 lb.	REMARKS.
12 x 18	24	Pittsburgh Forge & Iron Co. Allegheny, Pa. (1888)	56½ in.	44 ft.	158 ft.	3 cars	22,000 lb.	50,000 lb.	108 tons	Sometimes hauls 4 cars=144 tons. Grade 125 feet long. Curve of 44 feet radius is ¼ circle and on level track. 900 lbs. coal fuel per day of ten hours.
12 x 18	24	Pennsylvania Tube Works... Pittsburgh, Pa. (1889)	56½ in.	80 ft.	176 ft.	1 car	22,000 lb.	40,000 lb.	31 tons	Has hauled 2 loaded cars=72 tons. Grade about 200 feet long and coming on curve of 138 feet radius. Besides shifting cars. this locomotive does the work of 11 laborers loading cars. A car is loaded in 15 minutes by pulling gas pipe up the skids onto the car by a rope attached to the locomotive. No repairs, except 4 cylinder cocks, for 3 years. Burns daily 1,140 lbs. slack coal fuel costing 15 cents.
12 x 18	24	Ohio Iron Co...... Zanesville, O. (1875)	56½ in.	115 ft.	185 ft.	4 cars	18,000 lb.	22,000 lb.	80 tons	Curves on level. Grades 300 and 200 feet long with only short, level approach. By taking a run has hauled 168 tons up 185 feet, and 84 tons up 240 feet grade. 900 lbs. coal fuel and 1 tank water per day.
		Same on different track......	56½ in.	115 ft.	240 ft.	3 cars	18,000 lb.	22,000 lb.	60 tons	
12 x 18	21	Woodward Iron Co.... Wheeling, Ala. (1884)	60 in.	318 ft.	185 ft.	6 cars	14,000 lb	empty	42 tons	Length of road 6 miles. Locomotive used for yard work.
		Same on different track......	60 in.	......	132 ft.	13 cars	22,000 lb.	40,000 lb	108 tons	
12 x 18	24	Cambria Iron Co......... (1879) Johnstown, Pa.	57 in.	127 ft.	185 ft.	3 cars	17,500 lb.	22,000 lb.	50 tons	Has hauled 4 cars 79 tons. About 100 miles, and 1,200 lbs. coal fuel daily.

Cylinders		Owner, location, and date	Gauge	Curve	Grade	Cars	Weight per car	Weight	Total	Remarks
18 x 18	24	Walter A. Wood Mowing & Reaping Machine Works, Hoosick Falls, N.Y. (1884)	56½ in.	150 ft.	185 ft.	3 cars	20,000 lb.	24,000 lb.	66 tons	Curve and grade come together. Grade 200 feet long, 600 lbs. coal fuel, 2 tanks of water per day of 10 hours. On nearly straight track, by taking a run, put 6 cars weighing 132 tons up grade of 185 tons per mile.
*12 x 18	24	Crystal Plate Glass Co, Crystal City, Mo. (1881)	50¾ in.	250 ft.	212 ft.	3 cars	18,000 lb.	24,000 lb	63 tons	84 miles, 2,300 lbs. coal fuel daily. The 250 feet curve comes on 212 feet grade.
12 x 18	24	Blair Iron & Coal Co., Holidaysburg, Pa. (1900)	57 in.	212 ft.	230 ft.	4 cars	8,000 lb.	15,000 lb.	30 tons	About 60 miles, 1,500 lbs. coal fuel daily.
*12 x 18	24	Eliza Furnace, Pittsburgh, Pa. (1873)	56½ in.	114 ft.	303 ft.	2 cars	18,000 lb.	24,000 lb.	42 tons	Grade 190 feet long, with level approach only long enough to hold engine and cars.
*13 x 20	24	P. & B. C. C. & I. Co, Uraina, Pa. (1875)	56¼ in.	360 ft.	132 ft.	7 cars	18,000 lb.	22,000 lb.	140 tons	60 miles daily, burning 760 lbs. coal fuel, and using 2½ tanks of water per day. Has hauled 17 empty cars—153 tons up 237 feet per mile. Railroad since abandoned.
14 x 20	24	Carnegie, Phipps & Co. (Ltd.), Pittsburgh, Pa. (1886)	57 in.	118 ft.	15 ft.	4 cars	20,000 lb.	40,000 lb.	120 tons	Curve over ¼ circle. On main track shifts trains that stall large shifting engines. Also, occasionally go around ¼ circle curve of 55 feet radius.
*14 x 20	24	Lucy Furnaces, Pittsburgh, Pa. (1880)	57 in.	90 ft.	78 ft.	20 cars	20,000 lb.	40,000 lb.	600 tons	Has hauled 26 cars. Has a short run. Grade short, with reversed curves. 90 feet curve on cinder siding. Handle 200 cars, run 80 to 100 miles, burn 2,300 lbs. coal fuel daily.
14 x 20	24	W. C. Allison, Philadelphia, Pa. (1881)	57 in.	......	96 ft.	10 cars	20,000 lb.	24,000 lb.	220 tons	500 lbs. coal fuel daily. Not worked to full capacity.
14 x 20	24	North Chicago Rolling Mill Co., South Chicago, Ill. (1884)	57 in.	200 ft.	200 ft.	3	cars fluid metal	cars fluid metal	45 tons	Usual work carrying fluid metal from furnace to converter. On special trial pushed 200 tons up 200 feet grade. Length of grade about 1,000 feet.
		Same on different track..	......	125 ft. level		6 cars / 41 cars	loaded 24,000 lb. / empty 24,000 lb.	40,000 lb.	186 tons / 1,160 tons	9° reversed curves on grade.
		Same on different track..	......	105 ft.		24 cars	24,000 lb.	17,000 lb.	228 tons	
14 x 24	24	Pennsylvania Steel Co., Steelton, Pa. (1884)	57 in.	105 ft.	......	8 cars	17,000 lb.	40,000 lb.	228 tons	1,680 lbs. anthracite fuel, 2 tanks of water daily. This locomotive is also used on curves of 53 ft. radius at another track.
14 x 24	24	Ohio Iron Co., Zanesville, O. (1888)	57 in.	115 ft.	240 ft.	8 cars	24,000 lb.	40,000 lb.	256 tons	1,200 lbs. coal fuel and 2 tanks water per day of 12 hours.

* Change in owner, or location, or service, since report was made.

Memoranda of Work done by our Special Service Locomotives, from Reports furnished by Owners.

Size of cylinders.	Page showing style.	Owner and Location, and Date of Report.	Gauge of track in inches.	Weight of rail in pounds per yard.	Length of road in miles.	Radius of sharpest curve in feet.	Grade in feet per mile.	Number of cars hauled at one time.	Weight of each car in pounds.	Load on each car in pounds.	Weight of train in tons of 2,000 lb.	REMARKS.
5 x 10	40	Pennsylvania Steel Co. Rail Mill. Steelton, Pa. (1884)	36 in.	......	375 ft.	none	level	1 car	1 200 lb.	5,400 lb.	3 tons	Track all inside of rail mill. About 21 miles, 475 lbs. anthracite, 4 tanks water daily. Locomotive hauls about 195 tons steel blooms daily.
5 x 10	38	H. Floersheim. Finleyville, Pa. (1889) Coal Mines.	40 in.	16 lb.	1 ml.	......	57 ft.	12 cars	800 lb.	3,000 lb.	23 tons	Sometimes hauls 17 cars=32 tons. Gets out about 600 tons coal daily.
5 x 10	36	E. J. Gaynor. Contractor. Latrobe, Pa. (1889)	24 in.	24 lb.	½ ml.	......	105 ft.	8 cars	950 lb.	empty	4 tons	Round trip every 7 minutes. Also on different track hauled 8 loaded cars=19 tons up grades of 75 feet per mile.
5 x 10	40	Oliver Bros. & Phillips. Pittsburgh. Clapp-Griffith Steel Works. (1888)	27 in.	50 lb.	200 ft.	158 ft.		1 car	6,000 lb.	3,500 lb.	5 tons	228 lbs. coal fuel per 12 hours. Runs day and night. Hauls hot ingots on solid cast-iron car with loose wheels and excessive friction.
5 x 10	40	Midvale Steel Works. Philadelphia, Pa. (1888)	36 in.	56 lb.	......	15 ft.	528 ft.	5 cars	2,300 lb.	empty	11 tons	10 per cent. grade 100 ft. long with level approach. Hauls trains of 21 tons on level. 175 lbs. coal fuel and 2 tanks of water daily.
6 x 10	41	Carrie Furnace. Pittsburgh. Coke Ovens. (1888)	56½ in.	54 lb.	¼ ml.	......	almost level	2 cars	3,000 lb.	9,750 lb.	13 tons	Usual work one coke larry. Charges 50 ovens daily. One tank of water and 400 lbs. slack coal fuel per 8 hours.
6 x 10	26	U. S. Engineers. Harbor Improvement. (1888) Yaquina, Oregon.	36 in.	25 lb.	1 ml.	200 ft.	15 ft.	10 cars	2,500 lb.	4,000 lb.	32½ tons	Full power not tested. 28 to 40 miles per day, burning about 400 lbs. coal fuel.

	No.	Owner or location	Gauge	Rail	Length	Curve	Grade				Weight	Remarks
x 10	38	Ferrocarril Ceiba........ South America. (1877)	38 in.	30 lb.	70 mls.	....	about 15 ft.	0	coffee	cars	60 tons	Used for shifting.
x 10	26	C. C. Pinckney, Jr........... Phosphate mines, Charleston, S. C. (1878)	49 in.	16 lb.	2 mls.	110 ft.	22 ft.	12 cars	2,000 lb.	5,000 lb.	42 tons	Full power not tested. 40 to 50 miles per day.
x 10	26	Arcata & Mad River R. R. .. Arcata, Cal. (1882)	45 in.	30 lb.	8 mls.	409 ft.	90 ft.	7 cars	2,000 lb.	8,000 lb.	35 tons	Has hauled 10 cars weighing 50 tons. 10 to 15 miles per hour speed. Grade 400 feet long. 64 to 80 miles, ¾ cord of wood fuel, 5 tanks of water daily.
x 10	26	C. C. Pinckney, Jr........ Phosphate mines, Charleston, S. C. (1879)	36 in.	16 lb.	2 mls.	225 ft.	94 ft.	10 cars	2,000 lb.	4,000 lb.	30 tons	Grade 1,300 feet long. Curve on grade. Has hauled 14 cars—42 tons. 37 to 62 miles, ¼ cord wood fuel daily. On trial hauled 8 loaded cars—28 tons up grade 2 miles in 6½ minutes.
x 10	41	Otis Iron & Steel Co........ Cleveland, O. (1886)	23 in.	40 lb.	2 mls.	40 ft.	99 ft.	2 cars	2,700 lb.	5,700 lb.	8 tons	Grade 70 ft. long. On test hauled 5 loaded cars—21 tons, starting on 40 ft. radius. Also hauled 59 tons round reverse curves of 100 ft. radius. 2,000 lbs. coal fuel, 6 tanks of water per day of 24 hours.
x 10	35	Bass, Kroeigk & Co........... Sugar Plantation, San Domingo. (1886)	30 in.	16 lb	6 mls.	50 ft.	114 ft.	4 cars	4,000 lb.	4,000 lb.	16 tons	Has hauled 10 cars—40 tons. Empty cars go up 172 feet per mile grade. Burns ¼ cord of wood fuel, uses 3 to 4 tanks water, and hauls about 250 tons of sugar-cane per day of 11 to 18 hours.
x 10	26	Juniata Mining & M'f'g Co... Iron mines, Tyrone, Pa. (1884)	36 in.	16 lb.	1 ml.	50 ft.	150 ft.	15 cars	1,000 lb.	4,000 lb.	30 tons	Curve and grade come together. 90 to 100 miles, 1,000 lbs. coal fuel, 4 tanks water per day of 10 hours. About 12 miles per hour usual speed.
x 10	34	Lithgow Bros.............. Sugar Plantation, Puerto Plata, San Domingo. (1881)	26 in.	14 lb.	4½ mls.	30 ft.	158 ft.	12 cars	853 lb.	2,240 lb.	13 tons	40 miles, 300 lbs. anthracite fuel, tanks filled 6 times daily. "Far superior to English locomotives."

* Change in owner, or location, or service, since report was made.

Memoranda of Work done by our Special Service Locomotives, from Reports furnished by Owners.—*Continued.*

Size of cylinders.	Page showing style.	Owner and Location, and Date of Report.	Gauge of track in inches.	Weight of rail in pounds per yard.	Length of road in miles.	Radius of sharpest curve in feet.	Grade in feet per mile.	Number of cars hauled at one time.	Weight of each car in pounds.	Load on each car in pounds.	Weight of train in tons of 2,000 lb.	REMARKS.
6 x 10	34	Sucesion de Don J. Latimer. Sugar plantation. (1880) Porto Rico.	30 in.	16 lb.	1½ mls.	55 ft.	158 ft.	8 cars	750 lb.	2,250 lb.	12 tons	36 miles, 800 lbs. coal fuel daily.
6 x 10	35	Italia Sugar Plantation. (1888) San Domingo.	22¹³⁄₁₆ in.				158 ft.	2 cars	3,000 lb.	6,800 lb.	10 tons	Grade 100 yards long. Each car loaded with 4 hogsheads of molasses.
6 x 10	26	Eureka Mill. Silver mine. Empire City, Nev. (1888)	30 in.	16 lb.	2 mls.	400 ft.	180 ft.	6 cars	2,400 lb	empty	6 tons	Loaded train, 36 tons, comes down grade. Runs 36 miles, burning ¾ cord pine wood fuel and using 4 tanks of water per day of 9 hours.
6 x 10	38	LongfellowCopperMiningCo. Copper mine. (1881) Clifton, Arizona.	20 in.	16 lb.	4 mls.	136 ft.	211 ft.	7 cars	3,000 lb.	empty	10 tons	32 to 40 miles, ½ cord wood fuel daily.
*6 x 10	26	Parrott Iron Works. Limestone, etc. Greenwood Iron Works, N. Y. (1884)	32½ in.	12 to 15 lb.	1½ mls.	200 ft.	308 ft.	2 cars	2,200 lb.	5,400 lb.	7½ tons	Has hauled 3 cars, weighing 11 tons. 800 lbs. coal fuel, 5 tanks of water per day of 24 hours. Hauls about 120 tons daily.
6 x 10	40	Oliver Bros. & Phillips. Clapp-Griffith Steel Works. Pittsburgh. (1888)	27 in.	50 lb.		16 ft.	343 ft.	1 car	3,250 lb.	14,000 lb.	8½ tons	Sometimes hauls 2 cars—17 tons. Grade 150 feet long, with sharp curve each end. 16 feet curve is ¼ circle and on 3 per 100 grade. Cast-iron cars with loose wheels with excessive friction used for hauling billets.
7 x 12	26	Laurel Valley Plantation. Sugar plantation. (1881) Thibodeaux, La.	36 in.	21 lb.	2½ mls.	80 ft.	4 ft.	10 cars	700 lb.	4,500 lb.	28 tons	Never worked to full capacity.

	No.	Owner, location, etc.										Remarks
7 x 12	41	Carnegie, Phipps & Co.(Ltd.) Homestead Steel Works. Homestead, Pa. (1888)	30 in.	56 lb.	....	16½ ft.	almost level	1	car of	blooms	5 tons	Curves of 16½ and 18 feet radius ¼ circle. Cast iron car, loose wheels. Sometimes hauls more.
7 x 12	20	Excelsior Stone Co........ (1888) Quarry, Lemont, Ills.	50½ in.	40 lb. 1¼ mls.			5 ft.	1 car	18,000 lb.	40,000 lb.	29 tons	Has hauled 10 box cars on nearly level track. Have steep grades and sharp curves in quarry. Haul 600 tons stone daily, burning 1,200 lbs. coal fuel, and using 3 tanks of water per day of 10 hours. Can make 100 miles daily.
*7 x 12	26	Nelson Bennett, contractor.. Cascade Tunnel, (1888) Wash. Ter.	36 in.	25 lb. 1¼ mls.			10 ft.	15 cars	2,000 lb.	4,500 lb.	32½ tons	55 miles, burning 600 lbs. coke fuel, and using 8 tanks of water per day of 24 hours. Ran a year night and day, longest stop for repairs, one hour.
7 x 12	38	Pittsburgh Steel Casting Co. Steel Works. Pittsburgh, Pa. (1881)	50½ in.	56 lb.			about 20 ft.	2 cars	ingots and moulds		about 12 tons	Sometimes hauls 3 or 4 cars from converter to rolling mill.
7 x 12	41	H. C. Frick Coke Co..... Mt. Pleasant, Pa. (1888)	50½ in.	50 lb.			say 20 ft.	5 cars	7,500 lb.	15,200 lb.	57 tons	Each larry holds about 200 bushels coal.
7 x 12	41	Jones & Laughlins (Ltd.) Steel mill. Pittsburgh, Pa. (1888)	23 in.	30 lb.		14 ft.	about 20 ft.	3 cars	5,050 lb.	3,700 lb.	13 tons	Often hauls 6 cars. Also pushes with pole 2 loaded standard-gauge cars—60 tons, and has pushed 4 cars—about 100 tons. Curves of 14, 15 and 20 feet radius for ¼ circle.
7 x 12	38	Linden Steel Co...... Pittsburgh, Pa. (1888)	36 in.			20 ft.	about 20 ft.	2	cars of	blooms	13 tons	Curves of 20, 25, 28 and 30 feet radius. Hauled 13 tons up 5 feet per 100 grade about 16 feet long, situated between reversed 25 feet radius curves.
*7 x 12	20	Upper & O'Connor........ Railroad contractors. Rockford, Ills. (1888)	38 in.	67 lb. 1¼ mls.			28 ft.	15 cars	1,800 lb.	5,600 lb.	55½ tons	80 miles, burning 600 lbs. coal fuel and using 8 tanks of water per day of 10 hours.

* Change in owner, or location, or service, since report was made.

Memoranda of Work done by our Special Service Locomotives, from Reports furnished by Owners.—Continued.

Size of cylinders.	Page showing style.	Owner and Location, and Date of Report.	Gauge of track in inches.	Weight of rail in pounds per yard.	Length of road in miles.	Radius of sharpest curve in feet.	Grade in feet per mile.	Number of cars hauled at one time.	Weight of each car in pounds.	Load on each car in pounds.	Weight of train in tons of 2,000 lbs.	REMARKS.
7 x 12	26	North Chicago Rolling Mill. South Chicago, Ills. (1884)	36 in.			60 ft.	say 28 ft.	2	cars hot ing ots		15 tons	Two of these locomotives are used on 36 inches track for hauling ingots, and two on 56½ inches track for miscellaneous work about blast furnaces and Bessemer converters. 75 tons is frequently hauled on level. Full power not tested.
*7 x 12	26	Steese Coal Co Youngstown, O. (1873)	39 in.		3 mls.		28½ ft.	30 cars	1,100 lb.	2,600 lb.	55 tons	Has hauled 40 cars. Gets out 200 tons coal per day.
7 x 12	26	J. C. Fishburne. Phosphate mines. Pon Pon, S. C. (1884)	36 in.	old 50 lb.	2 mls.	50 ft.	30 ft.	6 cars	about 3,000 lb.	6,800 lb.	27 tons	Has hauled 10 cars weighing 49 tons. 32 to 36 miles, 1 cord wood fuel, 5 tanks water per day of 11 hours.
*7 x 12	26	Massillon City Coal Co Massillon, Ohio. (1877)	39 in.	28 lb.	2½ mls.		37 ft.	40 cars	950 lb.	2,400 lb.	67 tons	Has hauled 52, and could haul 60 cars. Usual mileage 40, and best 50 miles per day. 800 lbs. coal fuel, and 4 tanks water per day. Does the work of 25 mules, and could get out 500 tons per day.
*7 x 12	28	Nelson Bennett, contractor. Cascade Tunnel, (1886) Wash. Ter.	36 in.	25 lb.	1½ mls.		35 ft.	15 cars	2,000 lb.	4,500 lb.	32½ tons	66 miles, burning 700 lb. coke fuel, and using 10 tanks of water per day of 24 hours. In constant use, night and day.

												Remarks
7 x 12	26	Fairmont Coal & Iron Co. Coal mines, Fairmount City, Pa. (1884)	36 in.	16 lb.	½ ml.	300 ft.	37 ft.	30 cars	1,300 lb.	empty	30 tons	Has hauled 52 cars weighing 34 tons. Cars come down grade loaded with 3,400 lbs. coal each. About 10 miles, 600 lbs. coal fuel, 3 tanks of water daily; gets out 800 tons coal daily.
7 x 12	41	Brown & Cochran, Coke works, Broadford, Pa. (1888)	56½ in.	60 lb.	¾ ml.	47 ft.	43 ft.	3 cars	4,000 lb.	8,000 lb.	18 tons	The 3 empty larries are hauled up grade of 132 feet per mile. The locomotive charges 127 ovens, and burns 300 lbs. coal fuel daily. No repairs for one year's running.
7 x 12	26	St. Mary's Coal Co., St. Mary's, Pa. (1884)	30 in.	25 lb.	1 ml.	39 ft.	45 ft.	50 cars	500 lb.	1,000 lb.	37 tons	Usual work 26 miles, 800 lbs. coal fuel, 2 tanks of water per day of 7 hours; has doubled this mileage. Gets out 500 tons coal daily, and does the work of 30 mules.
7 x 12	26	Julian Fishburne, Phosphate mines, S. C. (1884)	38½ in.	old 50 lb.	5 mls.	......	50 ft.	9 cars	1,200 lb.	6,720 lb.	36 tons	Has hauled 10 cars. 1,000 lbs. coal fuel, 2 tanks of water daily.
7 x 12	41	Mount Carbon Co. (Ltd.), Coke works, Powellton, W. Va. (1888)	56½ in.	56 lb.	......	......	50 ft.	2 cars	8,000 lb.	15,000 lb.	11½ tons	67 round trips—25 miles daily, charging 202 coke ovens on 72 hour charges with 2 larries. Never worked to full power. Grade 2,000 feet long.
7 x 12	26	Rose Mining & M'f'g Co., Phosphate mines, Charleston, S. C. (1888)	36 in.	30 lb.	2½ mls.	......	53 ft.	13 cars	about 2,000 lb.	4,000 lb.	39 tons	Grade 450 feet long. 60 miles per day of 12 hours, burning 1¾ cord wood fuel, and using 9 tanks of water.
7 x 12	37	Paint Creek R. R., Paint Creek, W. Va. (1884)	36 in.	25 lb.	5 mls.	179 ft.	53 ft.	10 cars	5,600 lb.	empty	98 tons	Has hauled 20 cars weighing 56 tons. Regular speed 10, and best 20 miles per hour. 60 to 80 miles, 2,000 lbs. coal fuel, 5 tanks of water per day of 10 hours. Gets out about 200 tons of coal daily, and could get out about 750 tons.
7 x 12	26	Eliza Furnace, Pittsburgh, Pa. (1881)	40 in.	56 lb.	½ ml.	24 ft.	78 ft.	3 cars	5,000 lb.	4,000 lb.	18 tons	Used for hauling hot cinder.

* Change in owner, or location, or service, since report was made.

Size of cylinders.	Page showing style.	Owner and Location, and Date of Report.	Gauge of track in inches.	Weight of rail in pounds per yard.	Length of road in miles.	Radius of sharpest curve in feet.	Grade in feet per mile.	Number of cars hauled at one time.	Weight of each car in pounds.	Load on each car in pounds.	Weight of train in tons of 2,000 lb.	REMARKS.
*7 x 12	26	Patterson & Kuhn, contract's Laurel Hill Tunnel, Pa. (1884)	36 in.	25 lb.	1 ml.	229 ft.	80 ft.	12 cars	1,800 lb.	empty	11 tons	Has hauled 21 cars weighing 22 tons. About 66 miles per 10 hours. Work night and day.
7 x 12	35	Aruba Phosphaat Maatschappy Phosphate mines. Aruba, S. A. (1884)	24 in.	18 lb.	3½ mls.	231 ft.	84 ft.	20 cars	1,600 lb.	2,000 lb.	36 tons	42 to 54 miles, about 560 lbs. coal fuel, and tanks filled 7 times daily.
*7 x 12	26	Parker & Karns City R. R. Parker, Pa. (1881)	36 in.	30 lb.	10 mls.	......	96 ft.	1	pass enger		10 tons	Usual work, shifting. Ran 10 miles up grade without train in 28 minutes.
7 x 12	41	H. C. Frick Coke Co. Broadford, Pa. (1888)	50½ in.	56 lb.	1 ml.	239 ft.	105 ft.	3 cars	7,500 lb.	15,200 lb.	34 tons	Hauling 3 larries, charges about 150 ovens daily.
*7 x 12	26	Belmont Coal & R. R. Co. Boyd's Switch, Ala. (1880)	33⅜ in.	16 lb.	2½ mls.	90 ft.	105 ft.	12 cars	2,200 lb.	empty	13 tons	Can haul 20 cars.
7 x 12	26	Conglomerate Mining Co. Iron mine. Delaware Mines, Mich. (1884)	36 in.	......	¼ ml.	130 ft.	106 ft.	3 cars	6,020 lb.	7,000 lb.	19 tons	During 8 months has handled 291,218 tons of cars and loads, burned 110,364 lbs. coal fuel, and run 4,180 miles.
7 x 12	26	Grist & Graham. Quarry. Lowellville, O. (1888)	36 in.	25 lb.	1½ mls.	......	105 ft.	6 cars	1,700 lb.	8,000 lb.	29 tons	Empty train=5 tons goes up grade of 297 feet per mile. Grades 500 and 600 feet long. 60 miles per day of 10 hours, burning 700 lbs. coal fuel, and using 6 tanks of water.

7 x 12	26	New River Mineral Co...... Ivanhoe Furnace, Va. (1838)	36 in.	25 lb.	1 ml.	90 ft.	119 ft.	3 cars	4,200 lb.	11,200 lb.	23 tons	Has hauled 4 cars—31 tons. 50 to 54 trips—100 to 108 miles, burning 1,800 lbs. of coal fuel, and using 5 tanks of water per day of 10 hours.
7 x 12	26	Brown, Bonnell & Co........ Iron mill, Youngstown, O. (1888)	32 in.	56 lb.	8 mls.	60 ft.	120 ft.	2 cars	{2,100 to 2,300 lb.	{6,000 to 18,000 lb}	about 15 tons	Has hauled on nearly level grade 6 cars—about 50 tons.
7 x 12	26	Fairmount Coal & Iron Co... New Bethlehem, Pa. (1887)	36 in.		1 ml.	......	132 ft.	35 cars	800 lb.	empty	14 tons	Grade 3,500 feet long. Gets out 1,000 tons coal daily. Hauled 15 loaded cars on trial—25 tons up grade. Ran 1 mile up grade in 2½ minutes.
7 x 12	20	Ch. H. Strong & Son......... Contractors. Cleveland, O. (1888)	36 in.				158 ft.	7 cars	1,400 lb.	6,200 lb.	27 tons	Short run approaching grade.
7 x 12	41	Pittsburgh & Connellsville Gas Coal Co......... Coke ovens. Scottdale, Pa. (1888)	50½ in.	56 lb.	¾ ml.		132 ft.	8 cars	4,800 lb.	11,200 lb.	24 tons	Charges 148 coke ovens, hauling 3 larries, burning 304 lbs. coal fuel, and using 2 tanks of water per day of 8 hours.
*7 x 12	26	Youngstown Coal Co........ Massillon, Ohio. (1979)	39 in.	28 lb.	1 ml.		150 ft.	25 cars	1,000 lb.	empty	12 tons	Has hauled 50 cars. Makes 16 trips, getting out 400 tons coal per day.
*7 x 12	27	Rusk Transportation Co...... Rusk, Tex. (1875)	36 in.	wood 4 x 4	15½ mls	360 ft.	156 ft.	2 cars	2,500 lb.	5,000 lb.	7½ tons	31 miles, burning 1½ cord wood fuel, and using 3 tanks of water per 6 hours. Road since changed to standard gauge.
7 x 12	41	New River Coke Co......... Coke ovens. Caperton, W. Va. (1888)	44 in.	30 lb.	¾ ml.	60 ft.	158 ft.	5 cars	5,200 lb.	empty	18 tons	Grade 900 feet long. Charges about 75 coke ovens in about 8 hours, burning about 140 lbs. coal fuel, and using 3 tanks of water.
7 x 12	26	Laughlin & Co............. (1881) Limestone quarry.	40 in.	20 lb.	2 mls.		158 ft.	6 cars	1,000 lb.	empty	8 tons	Never worked to full capacity.
*7 x 12	26	Brown & Mosgrove......... Kittanning, Pa. (1875)	42 in.	16 lb.	4 mls.	143 ft.	160 ft.	4 cars	3,000 lb.	6,500 lb.	19 tons	Can haul 6 cars. 480 lbs. coal fuel, 4 tanks water, 32 miles per 10 hours.

* Change in owner, or location, or service, since report was made.

Memoranda of Work done by our Special Service Locomotives, from Reports furnished by Owners.—Continued.

Size of cylinders.	Page showing style.	Owner and Location, and Date of Report.	Gauge of track in inches.	Weight of Rail in pounds per yard.	Length of road in miles.	Radius of sharpest curve in feet.	Grade in feet per mile.	Number of cars hauled at one time.	Weight of each car in pounds.	Load on each car in pounds.	Weight of train in tons of 2,000 lb.	REMARKS.
7 x 12	26	Crozer Steel & Iron Co...... Iron ore mines. Roanoke, Va. (1888)	36 in.	20 lb.	2 mls.	163 ft.	211 ft.	4 cars	4,000 lb.	empty	8 tons	Grade 1 mile long. Loaded train =32 tons comes down grade.
*7 x 12	26	H. W. Peters......... Coal mines. Fulton Landing, Ky. (1876)	57 in.	16 lb.	3 mls	85 ft.	220 ft.	8 cars	3,500 lb.	empty	14 tons	Grade ¼ mile long. Has hauled 8 cars. 450 lbs. coal fuel per day, 48 to 70 miles per day.
7 x 12	26	Helena & Livingston Smelting and Reduction Co...... Wickes, Mont. (1889)	30 in.	20 lb.	2¾ mls.	143 ft.	264 ft.	10 cars	1,700 lb.	empty	8½ tons	Grade 2 miles long, speed 7 miles per hour up grade.
7 x 12	26	S. Hamilton, Jr........ Croton Landing, N. Y. (1888)	49 in.				261 ft.	4	loaded	clay cars	12 tons	30 miles daily, burning 500 to 700 lbs. coal fuel.
7 x 12	38	Detroit Copper Mining Co... Clifton, Arizona. (1884)	20 in.	16 lb.	1¼ mls.	80 ft.	264 ft.	4 cars	3,300 lb.	6,000 lb.	19 tons	Also hauls 6 empty cars up grades of 220 feet per mile. 80 feet curve on 120 feet grade.
*7 x 12	26	Matilda Furnace...... Mt. Union, Pa. (1873)	27½ in.		3 mls.	171 ft.	270 ft.	5 cars	2,600 lb.	3,500 lb.	15 tons	Curve on grade. Short 351 feet grade.
7 x 12	27	Pennsboro & Harrisville R.R. Pennsboro, W. Va. (1876)	36 in.	20 lb.	8½ mls.	150 ft.	270 ft.	2 cars	{5,000 lb. 7,000 lb.}	{14,000 lb. empty}	13 tons	Grade ¾ mile long. 34 to 51 miles, 800 lbs. coal fuel per day. Hauls 20 tons up 300 feet grade. Wooden rail tried and abandoned.
8 x 12	35	Sugar Plantation...... Cuba. (1884)	30 in.	17 lb.	7 mls.		about 52 ft.	6 cars	about 3,000 lb.	7,000 lb.	36 tons	15 miles per hour usual speed. 56 miles, 1 cord wood fuel, 3 tanks water daily.

8 x 12	26	Ohio Valley R. R. & M. Co... De Koven, Ky. (1884)	44½ in.	30 lb.	1½ mls.	......	25 ft.	14 cars	2,700 lb	4,000 lb.	47 tons	Has hauled 20 cars weighing 67 tons. Regular speed 10 to 12, and best 15 miles per hour. 42 miles, 1,000 lbs coal fuel, 4 tanks of water per day of 10 hours. Usually hauls 375 to 450 tons coal per day; could haul 750 tons easily.
8 x 12	25	Rose Phosphate M. & M. Co. Charleston, S. C. (1884)	30 in.	30 lb.	2½ mls.	......	53 ft.	20 cars	1,000 lb.	4,000 lb.	50 tons	60 miles, 1½ cords wood, 6 tanks water daily.
*8 x 12	25	Shields & Dorwin, contract's. Elkton, Md. (1884)	30 in.	25 lb.	1 ml.	175 ft.	60 ft.	16 cars	1,000 lb.	6,000 lb.	56 tons	60 to 75 miles, 1,200 lbs. coal fuel, 3 tanks water per day of 10 hours.
8 x 12	37	East Florida Land & Produce Co. (Limited) St. Augustine, Fla. (1888)	30 in.	30 lb.	14 mls.	108 ft.	80 ft.	9 cars	5,000 lb.	5,000 lb.	45 tons	On return trip hauls empty train up 132 feet grade. Has hauled 58 tons up 80 feet grade, and 46 tons up 132 feet grade. 84 miles per day of 13 hours, burning ¾ cord wood and using 5 tanks of water. Usual speed 10 miles per hour with train, and fastest 26 miles per hour with one car.
*8 x 12	26	Keller & Reilly, contractors. Lancaster, Pa. (1884)	30 in.	old 50 lb.	2 mls.	......	150 ft.	14 cars	1,500 lb.	4,000 lb.	38 tons	40 miles, 800 lbs. coal fuel daily. Grade 500 feet long.
8 x 12	35	Santa Eulalia Silver Min. Co. Chihuahua, Mexico. (1887)	30 in.	20 lb.	9 mls	150 ft.	200 ft.	10 cars	1,500 lb.	empty	7½ tons.	Has hauled 12 cars—11 tons. Usual speed up grade 12 miles per hour, fastest on level with loaded train, 18 miles per hour. 36 miles, burning 600 lbs. coal fuel, and using 2 double tanks of water daily. This locomotive has both saddle and rear tank.
8 x 12	26	Crozer Steel & Iron Co. Iron ore mines. (1888) Roanoke, Va	30 in.	20 lb.	2 mls.	163 ft.	211 ft.	4 cars	4,000 lb.	empty	8 tons.	Sometimes hauls 5 cars. Grade 1 mile long. Cars come down grade loaded with 6 tons each.

* Change in owner, or location, or service, since report was made.

(119)

Memoranda of Work done by our Special Service Locomotives, from Reports furnished by Owners.—Continued.

Size of Cylinders.	Page showing style.	Owner and Location, and Date of Report.	Gauge of track in inches.	Weight of rail in pounds per yard.	Length of road in miles.	Radius of sharpest curve in feet.	Grade in feet per mile.	Number of cars hauled at one time.	Weight of each car in pounds.	Load on each car in pounds.	Weight of train in tons of 2,000 lb.	REMARKS.
8 x 11	40	Edgar Thomson Steel Works, Bessemer, Pa. (1881)	30 in.	60 lb.	......	27 ft.	slight	3	cars ingots		14 tons	Also hauls blooms in rail mill.
*8 x 14	26	Pioneer Coal Co., Kanawha Saline W. Va. (1875)	42 in.	25 lb.	1 ml.	35 ft.	12 ft.	30 cars	1,000 lb.	2,000 lb.	45 tons	Has hauled 42 cars. 10 bushels coal fuel, 3 tanks water, 10 trips per 8 hours. Can ship 1,800,000 bushels of coal per year.
*8 x 16	26	W. W. & Col. Riv. R. R., Walla Walla, W. T. (1875)	36 in.	25 lb.	32 mls.	......	44 ft.	9 cars	4,000 lb.	14,000 lb.	81 tons	Wooden rail tried and abandoned.
8 x 14	26	Longview Lime Works, Longview, Ala. (1888)	36 in.	24 lb.	3½ mls.	......	100 ft.	8 cars	2,000 lb.	9,000 lb.	44 tons	56 to 70 miles per day.
*8 x 14	26	Jackson Iron Co., Michigan. (1873)	56½ in.	28 lb.	6 mls.	152 ft.	154 ft.	3 cars	6,000 lb.	12,000 lb.	27 tons	48 miles per day.
*8 x 14	26	Knap & Co., Bloomfield Mines, Pa. (1873)	38 in.	35 lb.	1 ml.	200 ft.	154 ft.	16 cars	1,900 lb.	empty	15 tons	Has hauled 15 loaded cars—40 tons. About 100 miles daily, burning 600 lbs. coal fuel, and using ¼ tanks of water. Does work of 10 mules at cost of 1, besides saving in time.
8 x 14	26	Louisville Cement Co., Cement Rock Quarries, Louisville, Ky. (1873)	39 in.	20 lb.	2 mls.	......	154 ft.	14 cars	1,700 lb.	4,000 lb.	40 tons	

Cylinder	No.	Owner, location, or service	Gauge	Rail	Length	Grade	Grade	Cars	Load	Empty	Weight	Remarks
*8 x 14	20	Eureka Co........ Iron ore mines. Oxmoor, Ala. (1874)	36 in.	30 lb.	2½ mls.	....	185 ft.	8 cars	1,800 lb.	empty	8 tons	1½ cord wood fuel, 3 tanks water 6 to 9 trips per day. On trial hauled 43 tons up 132 feet grade.
8 x 14	20	Dade Coal Co...... Cole City, Ga.	36 in.	25 lb.	2½ mls.	....	185 ft.	15 cars	1,500 lb.	empty	11 tons	Usual work, switching. Has hauled 33 tons up same grade.
8 x 14	20	Blair Iron & Coal Co.... Furnace and mill. Roaring Springs, Pa. (1881)	50½ in.	45 lb.	⅛ ml.	....	185 ft.	2 cars	5,000 lb.	11,000 lb.	10 tons	700 lbs. coal fuel daily.
*8 x 14	20	Study & Co........ Iron ore mines. Huntingdon, Pa. (1881)	36 in.	60 lb.	4 mls.	50 ft.	224 ft.	2 cars / 7 cars	2,100 lb. / 2,400 lb.	8,990 lb. / empty	17 tons	Grade ¾ mile long, with 50 feet radius curves on grade. 48 to 56 miles, 990 lbs. coal fuel daily.
*8 x 11	28	Hawk's Nest Coal Co...... Ansted, W. Va. (1875)	30 in.	40 lb.	3½ mls.	112 ft.	318 ft.	20 cars	1,050 lb.	empty	10½ tons	2 miles of 240 feet grade. Has hauled 25 tons empty cars. Since this report this locomotive has been adapted to mine service, and now runs underground on easier grades.
*8 x 14	20	Soddy Coal Co. ...... Soddy, Tenn. (1880)	30½ in.	30 lb.	3 mls.	70 ft.	396 ft.	8 cars	2,000 lb.	empty	8 tons	Has hauled 12 cars—12 tons. 54 to 66 miles, 800 lbs. coal fuel daily.
8 x 16	20	Riverside Iron Works........ Furnace and mill. Wheeling, Va. (1881)	33 in.	30 lb.	1 ml.	75 ft.	53 ft.	5 cars	6,000 lb.	4,000 lb.	25 tons	Power not tested. Hauls hot cinder.
*9 x 14	20	Jno. R. Lee & Co...... R. R. contractors. Elgin, Ills. (1888)	38 in.	75 lb.	7 mls.		58 ft.	24 cars	1,400 lb.	0,000 lb.	88 tons	Has hauled 34 cars—109 tons. 120 to 140 miles, burning 1,000 lbs. coal, and using 4 tanks of water per day of 12 hours. Usual speed 12, and fastest 20 miles per hour.
*8 x 16	20	Sandersville & Tennille R. R. Sandersville, Ga. (1877)	60 in.	48 lb.	3¾ mls.	....	90 ft.	2 cars	baggage and passenger		18 tons	Has hauled 5 cars—75 tons. Usual speed 13, and best 25 miles per hour.

* Change in owner, or location, or service, since report was made.

(121)

Memoranda of Work done by our Special Service Locomotives, from Reports furnished by Owners.—*Continued.* (122)

Size of cylinders.	Page showing style.	Owner and Location, and Date of Report.	Gauge of track in inches.	Weight of rail in pounds per yard.	Length of road in miles.	Radius of sharpest curve in feet.	Grade in feet per mile.	Number of cars hauled at one time.	Weight of each car in pounds.	Load on each car in pounds.	Weight of train in tons of 2,000 lb.	Remarks.
*8 x 16	26	Natchez, Jackson & Col. R. R., Natchez, Miss. (1876)	42 in.	35 lb.	26 mls.	203 ft.	100 ft.	7 cars	4,000 lb.	11,000 lb.	52 tons	Curve on grade. 52 to 88 miles, 1½ cords wood fuel daily.
*8 x 16	26	Keystone C. & M. Co........ Coal mines. Meyersdale, Pa. (1875)	36 in.	30 lb.	5 mls.	....	100 ft.	10 cars	2,000 lb.	4,480 lb.	32 tons.	Sometimes hauls 12 cars. 1,680 lbs. coal fuel, 8 tanks water, 30 miles per 10 hours.
*8 x 16	26	Union Cons. Mining Co...... Copper mines. Ducktown, Tenn. (1875)	36 in.	30 lb.	5 mls.	....	105 ft.	8 cars	2,000 lb.	8,000 lb.	40 tons.	Has hauled more. 3¾ cord wood fuel, 3 tanks water, 30 miles per 10 hours.
9 x 14	41	H. C. Frick Coke Co......... Broadford, Pa. (1888)	56½ in.	56 lb	1 ml.	239 ft.	105 ft.	5 larries	7,500 lb.	15,200 lb.	57 tons	Each larry holds about 200 bushels coal.
9 x 14	26	Brown, Bonnell & Co........ Iron mill. Youngstown, O. (1888)	32 in.	56 lb.	3 mls.	60 ft.	120 ft.	2 cars	{12,100 to 12,700 lb.	6,000 to 20,000 lb.}	17 tons.	Has hauled 6 cars—about 55 tons.
*9 x 14	26	G. Peterson & Co., contract's, New reservoir. Washington, D. C. (1884)	35 in.	....	....	....	126 ft.	15 cars	2,000 lb.	8,000 lb.	75 tons.	3 cubic yards of clay in each car.
*8 x 16	26	H. B. Hays & Bro.......... Coal mines. Pittsburgh, Pa. (1880)	41 in.	30 lb.	¾ ml.	....	132 ft.	15 cars	1,100 lb.	2,000 lb.	20 tons.	Hauled 26 tons empty cars up same grade.

Cyl.		Owner and location	Gauge	Rail	Length			Cars	Load	Empty	Capacity	Remarks
9 x 14	26	Fairmount Coal & Iron Co. Coal mines. Fairmount City, Pa. (1884)	36 in.	30 lb.	3 mls.	205 ft.	132 ft.	40 cars	1,800 lb.	empty	20 tons	In winter haul 28 cars. 132 feet grade is 3,000 feet long, and rest of road is 90 feet per mile grade. 2 locomotives get out about 1,500 tons daily.
*8 x 16	26	Smith Mining Co...... Iron mines. Michigan. (1874)	57 in.				150 ft.	7 cars	6,000 lb.	14,000 lb.	70 tons	Grade 300 yards long. Took a run at it. Special test, usual work less, ran 5 miles in 12 minutes.
9 x 14	26	Frank Williams & Co...... Buffalo, N. Y. (1888)	50½ in.				176 ft.	1 car	20,000 lb.	50,000 lb.	35 tons	Grade 300 feet long on trestle.
8 x 16	26	Fairchance Furnace......... Iron ore and limestone. Fairchance, Pa. (1881)	36 in.	32 lb.	1½ mls.		180 ft.	12 cars	3,000 lb.	6,000 lb.	54 tons	24 to 36 miles, coal fuel daily.
9 x 14	26	Pittsburgh & Wheel. Coal Co. Bridgeport, O. (1884)	36 in.	40 lb.	¾ ml.	441 ft.	185 ft.	24 cars	900 lb.	empty	11 tons	Has hauled 45 cars weighing 20 tons. 40 miles, 2,000 lbs. coal fuel, 4 tanks water daily. Full capacity never tested.
*8 x 10	26	Kittle & Co., contractors .... Keokuk, Ia. (1874)	36 in.	28 lb.			193 ft.	10 cars	1,600 lb.	3,000 lb.	26 tons	Contractors' road, and track often shifted.
9 x 14	26	N. Chicago Rolling Mill Co... South Chicago, Ills. (1884)	50½ in.				200 ft.	1	car fluid	metal	13 tons	Generally used for moving 3 cars fluid metal, weighing 39 tons on level. On trial moved 18 cars of coke, weighing 450 tons, on level.
*8 x 16	26	Maple Grove Coal Co .... Raymilton, Pa. (1873)	56½ in.				210 ft.	3 cars	17,000 lb.	empty	27 tons	Has hauled 4 cars. Hauled over 300 tons on siding on slight grade.
9 x 14	26	Franklin Iron M'f'g Co..... Franklin Iron Works, N. Y. (1884)	36 in.		¼ ml.		225 ft.	1 car	14,000 lb.	24,000 lb.	18 tons	Used for hauling cinder from blast furnace.
*8 x 10	26	Contractors for Reservoir Water Works. Pittsburgh, Pa. (1875)	36 in.	25 lb.			247 ft.	7 cars	1,000 lb.	5,000 lb.	21 tons	Has hauled 27 tons.

* Change in owner, or location, or service, since report was made.

Memoranda of Work done by our Special Service Locomotives, from Reports furnished by Owners.—Continued.

Size of cylinders.	Page showing style.	Owner and Location, and Date of Report.	Gauge of track in inches.	Weight of rail in pounds per yard.	Length of road in miles.	Radius of sharpest curve in feet.	Grade in feet per mile.	Number of cars hauled at one time.	Weight of each car in pounds.	Load on each car in pounds.	Weight of train in tons of 2,000 lb.	REMARKS.
* 8 x 16	26	Miss. Val & Ship Island R. R. Vicksburg, Miss. (1881)	42½ in.	35 lb.	20 mls.	296 ft.	264 ft.	3 cars	6,000 lb.	12,000 lb.	27 tons	Grade on temporary track 400 feet long. Hauls 7 to 12 freight cars=63 to 108 tons over main line. 40 to 120 miles, 1,200 lbs. coal fuel daily. Usual speed 12 to 15 miles per hour, and best 20.
*9 x 14	26	Ryan & McDonald, contractors "West Shore," R. R. Newburg, N. Y. (1884)	38 in.	56 lb.	3 mls.	358 ft.	369 ft.	18 cars	1,800 lb.	empty	16 tons	About 150 miles per day of 22 hours. Moved 4,280 cubic yards excavated material ½ mile from steam shovel to dump, returning up grade of 7 per cent., 300 feet long, with empty cars in one day and night of 22 hours. Seven other locomotives doing similar work.
9 x 14	26	S. L. Brightbill & Son. Quarry, Annville, Pa. (1888)	56½ in.	56 lb.	2 mls.	....	370 ft.	3 cars	7,500 lb.	13,440 lb.	31 tons	Has hauled 4 cars=42 tons. Grade 300 feet long, 50 miles, burning 500 lbs. coal, and using 2 tanks of water daily.
9 x 14	26	Union Mining Co. Mt. Savage Firebrick Works. Mt. Savage, Md. (1888)	36 in.	30 lb.	2½ mls.	....	385 ft.	14 cars	1,450 lb.	empty	10 tons	Does the work of 7 mules in about half the time easily. Steepest grade 100 feet long. Rises 642 ft. in 2 miles.

9 x 14	20	Chestnut Hill Iron Ore Co... Columbia, Pa. (1884)	30 in.	60 lb.	½ ml.	65 ft.	528 ft.	2 cars	8,000 lb.	4,000 lb.	12 tons	16 miles, 1,000 lbs. coal fuel, 4 tanks of water each day of 24 hours, running day and night. Repairs of locomotive for 18 months, $50, during which time it hauled 47,000 tons of pig metal to shipping wharf, 150,000 tons of cinder, and ran about 8,700 miles. Cinder cars with end dump are used, and a steam cylinder and crane car attached to the locomotive, by which the cinder of furnaces making 600 tons iron per week is dumped at a cost of only $8 per day. Hauling by horse-power and dumping the usual way cost $23 per day. This apparatus is patented by Mr. Jerome L. Boyer, Reading, Pa., to whom we refer companies desiring to use it. (See page 66.)
9½ x 14	41	Carnegie, Phipps & Co. (Ltd.) Homestead Steel Works. Homestead, Pa. (1888)	30 in.	56 lb.	....	18 ft.	almost level	1 car	hot	ingots	7 tons	Occasionally hauls 4 cars—28 tons. Curve ¼ circle. Cast-iron cars, loose wheels, very hot, and excessive friction.
9½ x 14	21	Jos. Walton & Co... Elizabeth, Pa. (1888)	43½ in.	40 lb.	1½ mls.	95 ft.	95 ft.	30 cars	1,400 lb.	empty	20 tons	Sometimes hauls 40 cars—23 tons. Grade averages 80 feet per mile. 80 to 108 miles, burning 840 lbs. coal, using 5 tanks water, and getting out 30,000 bushels coal per day of 10 hours.

* Change in owner, or location, or service, since report was made.

Memoranda of Work done by our Mine Locomotives, from Reports furnished by Owners.

AND LOCATION, AND DATE OF REPORT.	Gauge of track in inches.	Weight of rail in pounds per yard.	Length of track in miles.	Radius of sharpest curve in feet.	Grade in feet per mile.	Number of cars hauled at one time.	Weight of each car in pounds.	Load on each car in pounds.	Weight of train in tons of 2,000 lbs.	REMARKS.
l Coal Co...... okfield, Ohio.	36 in.	16 lb.	1 ml.	......	105 ft.	10 cars	1,450 lb.	empty	7 tons	Grade 1,360 yards long in tunnel. Has hauled 20 cars. 600 lbs. coal fuel, 24 to 30 miles daily.
'd Coal Co.... ington, Ky.	44 in.	26 lb.	¾ ml.	50 ft.	slight	25 cars	1,100 lb.	2,900 lb.	50 tons	Sometimes hauls 37 cars=74 tons. Gets out 400 tons coal per day.
n & Son...... lenville, Ohio.	36 in.	27 lb.	½ ml.	40 ft.	48 ft.	37 cars	850 lb.	2,000 lb.	52 tons	Full power not tested.
& Co..... traitsville, Ohio.	42 in.	20 lb.	⅜ ml.	......	78 ft.	18 cars	1,220 lb.	4,000 lb.	47 tons	Has hauled 25 cars=65 tons. Does the work of 12 mules and 8 drivers. 45 miles, 500 lbs. coal fuel daily.
e Coal Co...... bridge, Ohio.	31 in.	30 lb.	1 ml.	150 ft.	105 ft.	20 cars	1,000 lb.	2,500 lb.	35 tons	Has hauled 27 cars weighing 47 tons. 28 to 32 miles, 1,000 lbs. coal fuel, 8 tanks water per day of 10 hours.
ill Coal Co...... ado, Wash. Ter.	36 in.	30 lb.	1¼ mls.	30 ft.	105 ft.	20 cars	1,500 lb.	empty	15 tons	Two locomotives get out about 800 tons coal daily.
oal Co.... nd City, W. Va.	42 in.	30 lb.	¼ mls.	40 ft.	105 ft.	15 cars	1,000 lb.	2,000 lb.	22½ tons	Has hauled 22 cars=33 tons.

7 x 12	30	Excello Coal & Mining Co ... Excello, Mo. (1888)	28 in.	18 lb.	1¾ mls.	......	142 ft.	16 cars	850 lb.	1,800 lb.	21 tons	Has hauled 38 cars—50 tons up grade of 44 feet per mile. 49 to 60 miles, burning 950 lbs. coal fuel, and using 5 tanks of water, getting out about 200 tons coal per day of 9 hours.
7 x 12	30	Kaul & Hall ... . St. Marys, Pa. (1890)	30 in.	16 lb.	1 ml.	144 ft.	158 ft.	40 cars	500 lb.	empty	10 tons	Curve comes on 105 feet grade. Has hauled 62 cars—15 tons. Gets out 400 tons coal daily.
7 x 12	30	Fairmount Coal & Iron Co... Fairmount City, Pa. (1884)	30 in.	18 lb.	¾ ml.	......	158 ft.	25 cars	1,300 lb.	empty	16 tons	Grade 400 feet long.
7 x 12	30	Great Kanawha Colliery Co... Coal Valley, W. Va. (1884)	31 in.	16 lb.	2 mls.	50 ft.	175 ft.	20 cars	1,000 lb.	empty	10 tons	Can get out 500 tons daily. 4 tanks water daily, and 2 tons coal fuel per week.
*7 x 12	30	Furnace Colliery... .... Pottsville, Pa. (1879)	40 in.	......	½ ml.	......	185 ft.	30 cars	3,000 lb.	empty	22 tons	
*7 x 12	30	M. Saxman, Jr., & Co. ...... Latrobe, Pa. (1876)	38 in.	28 lb.	½ ml.	30 ft.	185 ft.	5 cars	1,200 lb.	3,350 lb.	11 tons	Curve at foot of grade. Has hauled 10 cars—23 tons. 32 to 40 trips, getting out about 290 tons coal daily.
7 x 12	30	W. B. Brooks & Son. ...... Nelsonville, Ohio. (1878)	42 in.	26 lb.	......	40 ft.	220 ft.	20 cars	900 lb.	empty	9 tons	Generally hauls less ; has hauled more.

* Change in owner, or location, or service, since report was made.

Memoranda of Work done by our Mine Locomotives, from Reports furnished by Owners.—*Continued.*

Size of cylinders.	Page showing style.	OWNER AND LOCATION, AND DATE OF REPORT.	Gauge of track in inches.	Weight of rail in pounds per yard.	Length of track in miles.	Radius of sharpest curve in feet.	Grade in feet per mile.	Number of cars hauled at one time.	Weight of each car in pounds.	Load on each car in pounds.	Weight of train in tons of 2,000 lb.	REMARKS.
7 x 12	30	New Central Coal Co........ Lonaconing, Md. (1888)	42 in.	25 lb.	1⅛ mls.	56 ft.	220 ft.	13 cars	1,800 lb.	empty	11½ tons	Has hauled 15 cars=18½ tons. Also hauls 19 cars=17 tons up 130 feet per mile. 24 miles, burning 600 lbs. coal fuel, using 6 tanks water, and getting out 500 tons coal per day. On another track has also got out 500 tons per day, hauling 13 empty cars per trip 2 miles up average grade of 180 feet per mile.
*7 x 12	30	Fort Dodge Coal Co........ Carbon, Ia. (1876)	32 in.				270 ft.				6 tons	Has hauled 8 tons. Going down 270 feet grade hauls 6 tons with a rope up a grade of over 500 feet per mile.
7 x 12	30	Victor Coal Co., Limited .... Philipsburg, Pa. (1894)	42 in.	30 lb.	1 ml.	191 ft.	290 ft.	9 cars	1,700 lb.	empty	8 tons	Sometimes hauls 11 cars weighing 9 tons. 50 to 60 miles. 1,000 lbs. coal fuel, 15 tanks of water, getting out 562 tons of coal per day of 10 hours.
*7 x 12	30	N. Y. & O. Coal Co. .... (1874)	41 in.			45 ft.	385 ft.	4 cars	900 lb.	2,100 lb.	6 tons	Grade, 150 feet long with curve on it, and heavy grade before and after. Has hauled 7 cars.

*8 x 12	30	Stone Estate McKeesport, Pa. (1884)	38 in.	40 lb.	2 mls.	......	35 ft.	40 cars	1,200 lb.	3,900 lb.	91 tons	55 to 72 miles daily.
8 x 12	28	Knoxville Iron Co. Knoxville, Tenn. (1888)	32 in.	30 lb.	¾ ml.	......	192 ft.	16 cars	1,000 lb.	2,000 lb.	24 tons	Has hauled 23 cars—34½ tons. 36 to 40 miles, burning 800 lbs. coal fuel, and using 4 tanks of water per day of 9 hours.
8 x 12	28	Glen Mary C. & C. Co. Glen Mary, Tenn. (1884)	36 in.	20 lb.	¾ ml.	160 ft.	200 ft.	50 cars	600 lb.	empty	15 tons	100 feet curve comes on 150 feet grade. 300 feet grade is 3,000 feet long. Cars come down loaded with 1,400 lbs. each. 24 to 30 miles, 1,000 lbs. coal fuel, 1,200 gallons water daily, getting out 450 tons of coal.
*8 x 14	30	T. Longstreth Nelsonville, Ohio. (1877)	42 in.	30 lb.	⅓ ml.	50 ft.	92 ft.	15 cars	1,000 lb.	2,700 lb.	28 tons	Has hauled 25 cars, or 46 tons. 35 to 40 miles, 500 lbs. coal fuel. 4 tanks water per day. Does the work of 20 mules and 10 drivers, and could do the work of 30 mules and 15 drivers.
*8 x 14	30	U. C. & M. Co. Ottumwa, Iowa. (1873)	37¾ in.	35 lb.	1¼ mls.	......	105 ft.	30 cars	650 lb.	empty	12 tons	Also hauls 40 loaded cars—60 tons up a 78 feet grade. Usual work less.
8 x 14	30	Consolidation Coal Co. Mount Savage, Md. (1873)	36 in.	28 lb.	2½ mls.	35 ft.	130 ft.	30 cars	1,400 lb.	empty	21 tons	Road all underground. 50 miles, 1,000 lbs. coal fuel daily.
		Former work on different track.	36 in.	28 lb.	½ ml.	35 ft.	185 ft.	6 cars	1,200 lb.	4,480 lb.	17 tons	
*8 x 14	30	Cameron Coal Co. Cameron, Pa. (1878)	30 in.	18 lb.	2 mls.	350 ft.	158 ft.	25 cars	600 lb	1,100 lb.	21 tons	Rail too light. 32 to 48 miles daily. Has hauled 35 cars—40 tons.
*8 x 11	30	Avery Coal & Mining Co. Avery, Ia. (1882)	38 in.	30 lb.	¼ mls.	......	180 ft.	10 cars	650 lb.	2,000 lb	18 tons	50 to 65 miles daily. Ran 18 months without losing a trip.

* Change in owner, or location, or service since report was made.

Memoranda of Work done by our Mine Locomotives, from Reports furnished by Owners.—*Continued.*

Size of cylinders.	Page showing style.	Owner and Location, and Date of Report.	Gauge of track in inches.	Weight of rail in pounds per yard.	Length of track in miles.	Radius of sharpest curve in feet.	Grade in feet per mile.	Number of cars hauled at one time.	Weight of each car in pounds.	Load on each car in pounds.	Weight of train in tons of 2,000 lb.	REMARKS.
8 x 14	30	Westmoreland Coal Co... Irwins, Pa. (1875)	40 in.	24 lb.	1¾ mls.	40 ft.	185 ft.	24 cars	1,000 lb.	empty	12 tons	40 feet curve on 119 feet grade hardest work. Not fully tested. 40 miles, 500 lbs. coal fuel daily.
8 x 14	30	George's Creek Coal & Iron Co. Lonaconing, Md. (1873)	36 in.	28 lb.	⅜ ml.	....	265 ft.	10 cars	1,300 lb.	2,240 lb.	17 tons	Can haul more. 20 trips daily.
*8 x 16	30	S. M. Heaton & Co........ Raven Run, Pa. (1860)	48 in.	28 lb.	1½ mls.	45 ft.	53 ft.	16 cars	3,000 lb.	5,000 lb.	64 tons	Has hauled 25 cars=100 tons. Underground entry 1 mile long.
8 x 16	30	Newburgh Orrel Coal Co.... Newburgh, W. Va. (1877)	41½ in.	35 lb.	2½ mls.	....	61 ft.	14 cars	1,400 lb.	4,480 lb.	41 tons	40 miles, 640 lbs. coal fuel daily. Engine does the work of 20 mules and 10 drivers.
8¼ x 16	28	Consolidation Coal Co....... Mount Savage, Md. (1873)	36 in.	28 lb.	2½ mls.	35 ft.	185 ft.	40 cars	1,400 lb.	empty	28 tons	Road all underground. Grade 1,800 feet long. Has made 18 trips = 63 miles in 9 hours. Can haul more. 55 miles, 1,000 lbs. coal fuel daily.
*9 x 12	31	Susquehanna Coal Co........ Wilkes-Barre, Pa. (1875)	42 in.	35 lb.	1¼ mls.	127 ft.	137 ft.	22 cars	2,240 lb.	empty	25 tons	Has hauled 27 cars weighing 30 tons. Grade 600 feet long, and 127 feet radius curve comes on grade.

9 x 14	30	Beaver Creek & Cumberland River Coal Co Greenwood, Ky. (1888)	36 in.	....	3 mls.	191 ft.	78 ft.	20 cars	1,200 lb.	2,800 lb.	40 tons	Short 3 per cent. up grade for loaded cars overcome by momentum. Has hauled 42 cars. Empty trains return up 3 per cent. grades. Can get out 500 tons coal per day.
9 x 14	30	St. Bernard Coal Co. Earlington, Ky. (1888)	44 in.	30 lb.	1¼ mls.	90 ft.	118 ft.	20 cars	1,100 lb.	2,800 lb.	39 tons	Gets out about 450 tons coal daily. Not kept busy.
9 x 14	30	Beech Creek Cannel Coal Co. Peale, Pa. (1888)	36 in.	30 lb.	1 ml.	303 ft.	132 ft.	20 cars	850 lb.	2,850 lb.	37 tons	Has hauled 25 cars. 50 to 64 miles daily.
*9½x14	36	M. Graver & Co. Pittsburgh, Pa. (1864)	44 in.	56 lb.	1 ml.		32 ft.	18 cars	1,300 lb.	3,000 lb.	39 tons	Has hauled 24 cars weighing 51 tons. 28 to 30 miles daily.
9½ x 14	28	American Coal Co. Lonaconing, Md. (1878)	36 in.				422 ft.	16 cars	1,000 lb.	empty	8 tons	Experimental trial. The grade is long, all underground, and, as engine coming down slid with wheels locked, its use inside was abandoned, and it is now working outside on less grade.
10 x 14	30	Southwest Improvement Co. Pocahontas, Va. (1888)	36 in.	30 lb.	1 ml.	100 ft.	52 ft.	25 cars	2,600 lb.	4,480 lb.	88 tons	32 to 35 round trips daily. Has got out 1,872 tons coal in one day. Grade 900 feet long. Empty trains return up 2 per cent. grade ¾ mile long.
10 x 14	30	Southwest Improvement Co. Pocahontas, Va. (1888)	36 in.	40 lb.	2½ mls.	118 ft.	52 ft.	25 cars	2,600 lb.	4,480 lb.	88 tons	Has got out 825 tons in one day. Grade 1,200 feet long. Curves on sidings 64 feet radius
		Same on different road......	30 in.	40 lb.			105 ft.	30 cars	2,600 lb.	4,480 lb.	106 tons	Use 3 to 4 tanks of water running 80 miles.
10¼x16	28	Illinois Central C. & M. Co... St. John's, Ill. (1875)	36 in.	30 lb.	1¼ mls.		204 ft.	23 cars	030 lb.	2,240 lb.	33 tons	Special trial. Straight track.

* Change in owner, or location, or service, since report was made.

Memoranda of Work done by our Light Locomotives on Logging Railroads, from Reports furnished by Owners. (132)

Size of cylinders.	Page showing style.	Owner and Location, and Date of Report.	Gauge of track in inches.	Weight of rail in pounds per yard.	Length of road in miles.	Radius of sharpest curve in feet.	Grade in feet per mile.	Number of cars hauled at one time.	Weight of each car in pounds.	Load on each car in pounds.	Weight of train in tons of 2,000 lbs.	REMARKS.
5 x 10	34	Hampton & Miller, Edgerly, La. (1884)	36 in.	20 lb.	2 mls.	600 ft.	16 ft.	2 cars	3,000 lb.	6,250 lb.	9 tons	Has hauled 14 tons. Usual speed 10 miles per hour. 40 miles, 1 cord wood fuel, 12 tanks of water per day of 11 hours, getting out 25,000 feet of logs daily.
*6 x 10	34	A. W. Taylor, Summerville, S. C. (1881)	35½ in.	wood 4 x 4	4 mls.	200 ft.	15 ft.	4 cars	2,000 lb.	8,000 lb.	20 tons	40 to 48 miles, 1½ cords wood fuel daily. Engine does the work of 20 mules. Hauled 700,000 feet of logs in 2 months.
6 x 10	34	George S. Young, Myersdale, Pa. (1884)	36 in.	16 lb.	6 mls.	200 ft.	178 ft.	12 cars	2,000 lb.	empty	12 tons	48 to 60 miles daily, including 1 night trip, hauling bark. Hauls 150,000 feet of logs and 100 cords of bark weekly.
7 x 12	26	Roanoke Lumber & R. R. Co. Norfolk, Va. (1888)	42 in.	25 lb.	7 mls.	....	10 ft.	6 cars	4,000 lb.	10,000 lb.	42 tons	Gets out 15,000 to 25,000 feet of logs, running 56 to 84 miles, burning 2 cords wood fuel, and using 8 tanks of water per day of 12 hours.
7 x 12	26	Middlebrook & Bro. (1888) Plank, Tex.	36 in.	16 lb.	5 mls.	....	18 ft.	5 cars	3,500 lb.	25,000 lb.	71 tons	Has hauled 9 cars=90 tons. Usual speed 12, and fastest 20 miles per hour. 5 trips per day, hauling 50 to 60 logs per trip.

No.	Cylinders	Owner, location, date	Drivers	Rail	Length of road	Grade	Curve radius	Cars	Empty car	Loaded car	Load	Remarks
26	*7 x 12	Tram & Lumber Co., Beaumont, Tex. (1878)	36 in.	56 lb.	7 mls.		ab't 13	8 cars	3,500 lb.	9,000 lb.	50 tons	Hauls 800,000 feet logs monthly.
20	7 x 12	N. C. Foster, Fairchild, Wis. (1884)	56½ in.	20 lb.	13 mls.	100 ft.	40 ft.	6 cars	7,000 lb.	20,000 lb.	81 tons	52 to 78 miles, 4 cords wood fuel, 10 tanks of water daily. Usual speed 10 miles per hour. No special trial of speed or power made.
		Same on return trip.					290 ft.	6 cars	7,000 lb.	empty	21 tons	
20	*7 x 12	P. H. Ketcham, Averill, Mich. (1882)	36 in.		21 mls.		53 ft.	9 cars	4,250 lb.	11,750 lb.	72 tons	196 miles, getting out 210,000 feet of logs per day of 24 hours. Has made 298 miles, getting out 272,000 feet, in 24 hours. Hauled 9 cars 8 miles, unloaded train and returned in 33 min.
20	7 x 12	G. & J. Backus, Mecosta, Mich. (1881)	37 in.	16 lb.	5 mls.	1,146 ft.	40 ft.	9 cars	2,500 lb.	10,000 lb.	50 tons	70 miles, 1 cord wood fuel daily.
26	*7 x 12	Muscogee Lumber Co., Pensacola, Fla. (1875)	30 in.		3½ mls.		66 ft.	7 cars	2,700 lb.	8,000 lb.	37 tons	30 miles, ½ cord wood fuel daily. Now working on different track.
38	7 x 12	Isaac Conroe, Conroe's Switch, Tex. (1881)	36 in.	wood	3 mls.		80 ft.	4 cars	2,000 lb.	12,000 lb.	28 tons	60 miles daily.
38	7 x 12	Samuel Allen, Hartley, Tex. (1881)	36 in.	wood	3½ mls.	477 ft.	105 ft.	4 cars	1,200 lb.	5,000 lb.	12 tons	42 to 48 miles, 1 cord wood fuel daily.
26	7 x 12	Chippewa Lumber & Boom Co., Lumber Yard, Chippewa Falls, Wis. (1889)	26 in.			45 ft.	120 ft.	6 cars	1,500 lb.	7,000 lb.	21 tons	Often hauls 15,000 ft. of boards. Grade 600 ft. long with short approach.
39	7 x 12	H. Lienhard, Handsboro, Miss. (1884)	36 in.	16 lb.	2½ mls.	600 ft.	123 ft.	3 cars	4,900 lb.	12,000 lb.	25 tons	Can take up 5 cars—about 42 tons.
20	*7 x 12	Whitney & Remick, Edmore, Mich. (1883)	48 in.	20 lb.	5½ mls.		132 ft.	11 cars	3,000 lb.	empty	16 tons	65 miles, 1½ cords wood fuel daily. Hauled 20,120,059 feet in 117 days. Best week's haul, 1,328,317 feet.

* Change in owner, or location, or service, since report was made.

Memoranda of Work done by our Light Locomotives on Logging Railroads, from Reports furnished by Owners.—Continued. (134)

Size of cylinders.	Page showing style.	Owner and Location, and Date of Report.	Gauge of track in inches.	Weight of rail in pounds per yard.	Length of road in miles.	Radius of sharpest curve in feet.	Grade in feet per mile.	Number of cars hauled at one time.	Weight of each car in pounds.	Load on each car in pounds.	Weight of train in tons of 2,000 lbs.	REMARKS.
7 x 12	26	Thos. Keelor, Wetmore, Pa. (1888)	56½ in.	20 lb.	2½ mls.	260 ft.	150 ft.	1 car	4,000 lb.	25,000 lb.	14½ tons	320 feet rise in 2½ miles, and some short 7 per cent. grades. 55 miles daily, and has made round trip in ½ hour. Costs 25 cents per 1,000 to deliver lumber from mill to shipping point. Has hauled 17,000,000 feet at $12 cost for repairs.
7 x 12	26	D. K. Ramey & Co., Ramey, Pa. (1878)	42 in.	16 lb.	2 mls.	60 ft.	185 ft.	2 cars	2,800 lb.	13,000 lb.	16 tons	40 to 48 miles, 1,000 lbs. coal fuel daily. Engine does the work of 20 horses and 10 men.
7 x 12	20	Geo. B. Merrill & Co., Lock Haven, Pa. (1884)	56½ in.	25 lb.	6 mls.	300 ft.	210 ft.	12 cars	2,000 lb.	empty	12 tons	84 to 96 miles, 1,600 lbs. coal fuel, 7 tanks per day of 12 hours. Grade 600 ft. long. Has hauled regularly loaded train about 60 tons up grade of 39 ft. per mile. Can get out 125,000 feet daily. Ran 40 miles per hour.
7 x 12	26	Amos Kent, Tangipahoa, La. (1881)	36 in.	25 lb.	2 mls.	90 ft.	212 ft.	2 cars	2,500 lb.	8,000 lb.	10 tons	40 miles, ¼ cord wood fuel daily. Engine does the work of 15 mules and 7 men, and hauls weekly 116,000 feet of logs, and could haul more if not delayed for loading and unloading.
7 x 12	26	E. B. Eddy M'f'g Co., Ltd, Hull, Canada. (1888)	45 in.	35 lb.	1½ mls.	120 ft.	237 ft.	{ 4 cars / 6 cars	1,200 lb. / 1,200 lb.	6,000 lb. / empty }	18 tons	Also hauls 10 loaded cars—36 tons on easier grades. 30 to 36 miles, burning 432 lbs. anthra-

7 x 12	38	Lewis A. Davis & Bro.......... Orange Bluff, Fla. (1888)		36 in.	20 lb.	6 mls.		250 ft.	5 cars	2,000 lb.	12,000 lb.	35 tons	cite coal fuel, and using 6 to 7 tanks of water per day of 10 hours. Two locomotives are used for hauling lumber from the mills to the piling ground, and are equipped with a patent exhaust, preventing any throwing of sparks.
7 x 12	30	Hopkins & Weymouth.......... Lock Haven, Pa. (1888)		50½ in.	25 lb.	3 mls.		220 ft.	7 cars	1,000 lb.	4,000 lb.	17½ tons	Grade short, and surmounted by momentum. Empty train goes up grade of 316 feet per mile 600 feet long, 72 miles, burning ¾ cord of wood fuel, and using 3 tanks of water per day of 10 hours.
7 x 12	39	Wesson & Persons.......... Bogue Chitto, Miss. (1881)		36 in.	wood	4 mls.		447 ft.	1 car	3,000 lb.	12,000 lb.	7 tons	Grade 300 feet long. Hauls 14 cars—35 tons up grades of 105 feet per mile and 600 feet long, and divide train on 290 feet grade. Usual speed 9 miles per hour. 30 miles, burning 800 lbs. coal fuel, and using 6 tanks of water per day of 11 hours.
8 x 12	20	A. J. & C. E. Covell.......... Whitehall, Mich. (1884)		36 in.	20 lb.	6 mls.		40 ft.	6 cars	5,200 lb.	12,000 lb.	52 tons	Has hauled 2 cars—11 tons. Grade 180 feet long, 32 to 56 miles, ¾ cord wood fuel daily. Engine does the work of 24 mules. Hauls 144,000 feet lumber weekly, and could do more if not delayed for loading and unloading. On special trial ran 36 miles per hour.
		Same on return trip.......... (1884)						140 ft.	6 cars	5,200 lb.	empty	16 tons	Sometimes hauls 7 cars weighing 60 tons. Usual speed about 15 miles per hour. 72 to 96 miles, getting out 70,000 feet of logs per day of 10 hours. Road very crooked with reverse curves.

* Change in owner, or location, or service, since report was made.

Memoranda of Work done by our Light Locomotives on Logging Railroads, from Reports furnished by Owners.—Continued. (136)

Size of cylinders.	Page showing style.	Owner and Location, and Date of Report.	Gauge of track in inches.	Weight of rail in pounds per yard.	Length of road in miles.	Radius of sharpest curve in feet.	Grade in feet per mile.	Number of cars hauled at one time.	Weight of each car in pounds.	Load on each car in pounds.	Weight of train in tons of 2,000 lb.	REMARKS.
8 x 12	20	John Du Bois...... Du Bois, Pa. (1884)	56½ in.	16 lb.	3½ mls.	423 ft.	50 ft.	10 cars	3,300 lb.	5,500 lb.	44 tons	Often hauls 12 to 14 cars. 35 to 56 miles, 450 lbs. coal fuel, water tank filled up 5 times daily. Can get out 50,000 to 75,000 feet of logs daily. Grade 2,000 feet long.
*8 x 12	39	Louis Sands...... Manistee, Mich. (1884)	38 in.	25 lb.	4 mls.	....	....	8 cars	2,800 lb.	empty	11 tons	Grade not measured, probably 52 feet per mile. No trial of speed or power made. 64 miles per day of 11 hours, getting out 128,000 feet of logs daily.
* 8 x 12	37	Jackson, Whaley & Co...... Suffolk, Va.	42 in.	30 lb.	25 mls.	229 ft.	100 ft.	8 cars	4,500 lb.	12,000 lb.	66 tons	On trial ran 12 miles in 21 minutes. Two years' repairs not over $50.
8 x 12	20	Whiting & Denning...... Denning, W. Va. (1888)	56½ in.	20 lb.	6½ mls.	180 ft.	213 ft.	12 cars	1,750 lb.	empty	10½ tons	Cars carry 12,000 lbs. each coming down. Has short 408 feet per mile grade on branch road. 65 miles, burning 1,500 lbs. coal fuel, using 5 tanks of water, and getting out 100,000 feet of logs daily. Has got out 163,000 feet in one day; also hauled in 20 days 12,254 logs, scaling 2,002,058 feet. Usual speed 10 miles per hour with loaded, and 15 miles with empty train.

*8 x 16	27	G. E. Pritchett & Co Gourdin's, S. C. (1874)	60 in.	wood	18½ ml.		12 ft.	5 cars	5,000 lb.	10,000 lb.	35 tons	54 miles daily.
*8 x 14	20	Lake George & Muskegon River R. R. Farwell, Mich. (1877)	50½ in.	25 lb.	7½ mls.		33 ft.	15 cars	2,500 lb.	12,000 lb.	100 tons	Runs night and day. 180 to 210 miles, 4 cords soft pine wood fuel, 12 tanks water per 24 hours. Has made 6 trips with 15 cars in 12 hours, getting out 908 logs, scaling 150,000 feet. One of the locomotives for this railroad was run by its own steam 15 miles over dirt roads, with teams to carry water and help turning corners and passing over stumps.
*8 x 14	26	Norris & Uhl, Agents. Stanwood, Mich. (1878)	50½ in.	25 lb.	7 mls.	290 ft.	44 ft.	12 cars	3,000 lb.	16,000 lb.	101 tons	Has hauled 140 tons. 75 to 90 miles, 1¾ cords pine wood fuel daily.
*8 x 14	26	Milner. Caldwell & Flowers Lumber Co. Bolling, Ala. (1870)	36 in.	20 lb.	4 mls		53 ft.	7 cars	2,850 lb.	5,000 lb.	27 tons	Has hauled more. 48 miles, ¾ cord pine wood fuel daily.
*8 x 14	26	D. & E. Lehoeuf. Evart, Mich. (1878)	50½ in.	25 lb.	6 mls.		53 ft.	12 cars	3,500 lb.	14,000 lb.	104 tons	Has hauled 15 cars—140 tons. Has run 4 miles in 6 minutes.
*8 x 16	20	E. & C. Eldred. Evart, Mich. (1878)	36 in.	25 lb.	8½ mls.		53 ft.	6 cars	8,000 lb.	18,000 lb.	78 tons	110 to 163 miles per day of 12 hours. Has hauled 38,000,000 feet of logs in 9 months, running night and day.
		Same with empty train. (1878)					212 ft.	6 cars	8,000 lb.	empty	24 tons	
8 x 14	29	Calcasieu Lumber Co. Lake Charles, La. (1884)	36 in.	30 lb.	4 mls.	315 ft.	57 ft.	10 cars	4,000 lb.	12,000 lb.	80 tons	35 to 60 miles, 1 cord of wood fuel, 8 tanks of water per day of 10 hours, getting out 60,000 feet of yellow pine logs daily. Usually carry 100 logs each trip.

* Change in owner, or location, or service, since report was made.

Memoranda of Work done by our Light Locomotives on Logging Railroads, from Reports furnished by Owners.—*Continued.*   (188)

Size of cylinders.	Page showing style.	Owner and Location, and Date of Report.	Gauge of track in inches.	Weight of rail in pounds per yard.	Length of road in miles.	Radius of sharpest curve in feet.	Grade in feet per mile.	Number of cars hauled at one time.	Weight of each car in pounds.	Load on each car in pounds.	Weight of train in tons of 2,000 lb.	REMARKS.
8 x 14	11	Turner & Oates, Lumbarton, Ga. (1884)	60 in.	60 lb.	5 mls.	50 ft.	100 ft.	6 cars	10,000 lb.	20,000 lb.	90 tons	Grade 350 feet long. 60 to 70 miles, 1 cord wood fuel, 4 tanks of water, getting out 36,000 feet of logs daily. 56 logs carried in each train. Has run 35 miles per hour.
8 x 16	10	Cody & Moore, Lake City, Mich. (1884)	36 in.	30 lb.	7½ mls.	636 ft.	75 ft.	14 cars	3,500 lb.	14,000 lb.	122 tons	Usual speed 15, and best 30 miles per hour. 240 to 255 miles, 10 cords wood fuel, 10 tanks of water, getting out 300,000 feet of logs per day and night of 24 hours. Hauled 10,000,000 feet in 40 days, stopping only 12 hours on Sundays, and during this time lost only 2 hours for packing and repairing locomotive.
*8 x 16	26	J. J. Dale & Co, Savannah, Ga. (1874)	60 in.	strap rail	4 mls.	......	75 ft.	4	cars	lumber	22 tons	Sometimes hauls 6 cars=37 tons. ¾ cord pine wood fuel daily. On special test ran 1 mile in 2½ minutes with 1 passenger coach.
*8 x 16	20	G. L. & D. E. Wing, Evart, Mich. (1878)	56½ in.	25 lb.	7½ mls.	......	78 ft.	14 cars	3,000 lb.	18,000 lb.	147 tons	96 miles, 1⅛ cords wood fuel daily.

Cylinders		Owner, Location (Date)	Gauge	Rail	Length		Grade	Cars			Weight	Remarks
*8 x 16	26	White, Friant & Co., Blanchard, Mich. (1881)	54½ in.	25 lb.	7 mls.	521 ft.	80 ft.	10 cars	4,000 lb.	16,000 lb.	100 tons	108 to 112 miles, 3 cords wood fuel daily. Gets out 600,000 feet of logs weekly.
8 x 16	30	Smith Lumber Co., Kalkaska, Mich. (1888)	54½ in.	50 lb.	5 mls.	....	80 ft.	4 cars	20,000 lb.	25,000 lb.	90 tons	Has hauled 6 cars—125 to 150 tons. 50 to 60 miles per day, putting in 50,000 to 100,000 feet per day. Could put in 100,000 feet daily on 8 to 10 miles road.
8 x 16	26	Sulphur Lumber Co., Buchanan's, Tex. (1884)	36 in.	30 lb.	2 mls.	....	90 ft.	12 cars	1,000 lb.	8,000 lb.	54 tons	40 to 50 miles, ¾ cord wood fuel, 3 tanks water per day of 11 hours.
8 x 16	26	Sierra Nev. Wood & Lum. Co. Virginia, Nev. (1881)	36 in.	35 lb.	2 mls.	286 ft.	100 ft.	3 cars	4,800 lb.	20,000 lb.	37 tons	48 to 60 miles daily; hauls daily 200 cords of wood, and could haul double.
*8 x 16	27	James Beard & Co., Alcona, Mich. (1879)	42 in.	25 lb.	6 mls.	....	100 ft.	6 cars	2,000 lb.	14,000 lb.	48 tons	72 miles, 1,000 lbs. coal fuel daily.
8 x 16	26	W. H. Hyde & Co., Ridgway, Pa. (1884)	54½ in.	35 lb.	6 mls.	....	158 ft.	4 cars	5,500 lb.	7,000 lb.	25 tons	36 miles, 250 lbs. coal fuel, getting out 18,000 feet daily. Sharp curves.
8 x 16	20	J. B. Walker, Penfield, Pa. (1888)	54½ in.	......	4 mls.	....	240 ft.	8 cars	800 lb.	10,000 lb.	43 tons	Put in 7,000,000 feet of pine, 2,000,000 hemlock, and about 1,190 tons of bark in 7 months. This locomotive, on this and other tracks, has put in about 50,000,000 feet in 8 years.
*8 x 14	26	Wyman, Buswell & Co., Sand Lake, Mich. (1878)	54½ in.	25 lb.	4 mls.	....	250 ft.	8 cars	4,000 lb.	empty	16 tons	On trial hauled 6 loaded cars, weighing 48 tons, up 250 feet.
*8 x 16	26	Geo. E. Williams, Dagus Mines, Pa. (1883)	54½ in.	......	5 mls.	242 ft.	300 ft.	10 cars	800 lb.	6,000 lb.	34 tons	90 miles, 1,496 lbs. coal fuel, 3 tanks of water, getting out about 100,000 feet per day of 10 hours. Also hauls 3 loaded cars, weighing about 15 tons, up an incline 400 feet long with a grade of about 700 feet per mile, with a rope, the locomotive running up a grade of about 200 feet per mile. Ran 3 years without losing a trip.
		Same on return trip. (Grade 700 feet long.)					528 ft.	10 cars	800 lb.	empty	4 tons	

* Change in owner, or location, or service, since report was made.

Memoranda of Work done by our Light Locomotives on Logging Railroads, from Reports furnished by Owners.—Continued. (140)

Size of cylinders.	Page showing style.	Owner and Location, and Date of Report.	Gauge of track in inches.	Weight of rail in pounds per yard.	Length of road in miles.	Radius of sharpest curve in feet.	Grade in feet per mile.	Number of cars hauled at one time.	Weight of each car in pounds.	Load on each car in pounds.	Weight of train in tons of 2,000 lbs.	REMARKS.
*9 x 14	21	Poitevent & Favre, Pearlington, Miss. (1885)	36 in.	25 lb.	13 mls.		slight	16 cars	4,000 lb.	8,000 lb.	96 tons	Greatest speed 20 miles per hour. Greatest mileage 150 miles per day of 11 hours. Usual mileage 100 miles, putting in 30,000 to 50,000 feet of logs.
9 x 14	21	Desha Lumber & P. Co., Arkansas City, Arks. (1889)	56½ in.	30 lb.	4 mls.		slight.	8 cars	5,000 lb.	17,000 lb.	88 tons	Has hauled 13 cars with 25,000 ft. oak logs =about 145 tons of train.
9 x 14	22	Cummer Lumber Co., Cadillac, Mich. (1885)	36 in.	30 lb.	9 mls.		13 ft.	30 cars	4,900 lb.	14,800 lb.	235 tons	Time for 9-mile run 40 minutes, hauling 55,818 feet. Grade 1 mile long. Could haul more.
9 x 14	10	Olive & Sternenberg, Olive, Texas. (1884)	36 in.	50 lb.	4 mls.	500 ft.	65 ft.	6 cars	6,000 lb.	10,000 lb.	78 tons	Gets out 80,000 feet of logs daily.
9 x 14	22	Sierra Lumber Co., Red Bluff, Cal. (1883)	30 in.	30 lb.	10 mls.	100 ft.	80 ft.	6 cars	3,500 lb.	9,000 lb.	37 tons	Has hauled 10 cars weighing 62 tons. Grade ½ mile long. 212 feet grades returning with empty train. 42 miles, 1½ cords wood fuel, 2 tanks water per day of 12 hours, getting out 45,000 feet of logs daily.
		Same on different track. (1887)					212 ft.	8 cars	8,500 lb.	15,000 lb.	74 tons	
9 x 14	21	Sisson & Lilley Lumber Co., Lilley, Mich. (1886)	56½ in.	25 lb.	4 mls.		90 ft.	6 cars	5,000 lb.	16,000 lb.	63 tons	Empty train-15 tons goes up grade of 180 feet per mile.

0 x 14	22	J. C. Trullinger, Astoria, Oregon. (1886)	50½ in.	56 lb.	2 mls.		132 ft.	4 cars	10,000 lb.	30,000 lb.	80 tons	Some logs hauled are 8 feet diameter, and will average 4 feet diameter scaling 1,800 feet per log. Grade 600 feet long. Engine not used to full capacity, and not $3.00 repairs in 12 months' use.
0 x 14	16	J. S. & W. M. Rice, Hyatt, Tex. (1884)	36 in.	56 lb.	4 mls.		150 ft.	6 cars	7,000 lb.	empty	21 tons	On trial hauled 6 loaded cars—about 90 tons. 80 miles, 1 cord wood fuel, 2 tanks water daily. No repairs in 12 months.
9 x 14	22	Buffalo Lumber Co., Bayard, W. Va. (1888)	36 in.	30 lb.	4 mls.		175 ft.	4 cars	about 4,000 lb.	empty	8 tons	Have grade of 419 feet per mile, ½ mile long, on branch.
9 x 14	22	Wilson, Kistler & Co. Rolfe, Pa. (1888)	56½ in.	30 lb.		570 ft.	180 ft.	5 cars	20,000 lb.	empty	50 tons	Shifts 10 cars—350 to 400 tons on nearly level track. Engine burns 4½ tons bituminous coal fuel per month, and uses 2 tanks of water per day. Grade 900 feet long.
0 x 14	21	Thompson & Tucker Lum. Co. Trinity, Tex. (1888)	30 in.	30 lb.	¼ mls.	75 ft.	257 ft.	6 cars	5,000 lb.	empty	15 tons	Could haul twice as much. Puts in 40,000 to 50,000 feet of logs, working 8 hours actual time. Can supply 2 mills, say 90,000 feet per day, on 5 to 8-mile haul. No repairs in 10 months. Grade 900 feet long.
0 x 14	14	R. T. Hardesty, Bubbin, Tex. (1881)	30 in.	20 lb.	6 mls.	935 ft.	817 ft.	5 cars	5,000 lb.	empty	13 tons	One cord of wood fuel, 2 tanks water, gets out 18,000 feet of logs daily. Grade 800 feet long.
9½ x 14	22	State Lumber Co. Fletcher. Mich. (1888)	36 in.		13 mls.		not stated	15 cars	5,000 lb.	10,000 lb.	157 tons	For 6 weeks made an average of 133,000 feet per day, hauling 15 cars 4 trips.

* Change in owner, or location, or service, since report was made.

Memoranda of Work done by our Light Locomotives on Logging Railroads, from Reports furnished by Owners.—*Continued.* (142)

Size of cylinders.	Page showing style.	Owner and Location, and Date of Report.	Gauge of track in inches.	Weight of rail in pounds per yard.	Length of road in miles.	Radius of sharpest curve in feet.	Grade in feet per mile.	Number of cars hauled at one time.	Weight of each car in pounds.	Load on each car in pounds.	Weight of train in tons of 2,000 lb.	REMARKS.
*9½x14	16	Palmer, Nichols & Co., Greenville, Mich. (1884)	37 in.	.....	13 mls.	573 ft.	86 ft.	5 cars	6,000 lb.	12,000 lb.	45 tons	Has hauled 65 tons up 86 feet grades, and 31 loaded cars, weighing about 270 tons, on easy grades; also 14 empty cars, weighing 42 tons, up grade of 102 feet per mile, 90 to 110 miles, getting out about 45,000 feet of logs daily.
*9½x14	14	Tawas & Bay Co. R. R., Tawas, Mich. (1879)	38 in.	20 lb.	21 mls.	.....	105 ft.	20 cars	3,500 lb.	10,000 lb.	135 tons	Grade 1¼ mile long. Has hauled 40 cars, weighing 225 tons, up 53 feet per mile grade. Costs 50 cents per 1,000 feet to haul 21 miles.
9¼x14	14	Van Dorn & Smith, Gurdon, Ark. (1884)	36 in.	20 lb. 35 lb.	3 mls.	382 ft.	111 ft.	10 cars	5,000 lb.	15,000 lb.	100 tons	Grade 3,000 feet long, and speed of train approaching grade is increased to about 25 miles per hour in order to carry train up. About 32 miles, less than ¼ cord wood fuel, and 1 tank of water daily. Hauls 30,000 feet of logs and the lumber cut by 3 mills daily.
9½ x 14	21	Satsop Railroad, Shelton, Wash. Ter. (1888)	56½ in.	40 lb.	12 mls.	477 ft.	131 ft.	3 cars	5,600 lb.	12,000 lb.	26 tons	Grade 5,000 feet long. 100 miles per day. Usual speed 20 miles per hour.
*9½x14	14	Etowah & Deaton's R. R., Etowah, Ga. (1881)	36 in.	30 lb.	6 mls.	318 ft.	150 ft.	5 cars	8,000 lb.	20,000 lb.	70 tons	Has hauled 6 cars—90 tons, 85 to 95 miles, 2 cords pine wood fuel daily.

Cylinder	No.	Owner, location, date	Diameter	Rail	Miles	Radius	Grade	Cars	Load	Empty	Tons	Remarks
9½ x 14	16	Warren Lumber Co........ Warren, Tex. (1888)	96 in.	20 lb. to 50 lb.	5 mls.	500 ft.	160 ft.	4 cars	5,000 lb.	27,000 lb.	65 tons	Taking a run, hauls 7 cars. 114 tons up 160 feet grade. Curve of 500 feet radius comes on grade. Also hauls 10 heavily loaded cars—190 tons up grade of 60 feet per mile. Gets in daily 100,000 feet of logs on a 4½ mile haul.
*9½x14	14	Smith, Taft & Marbury..... Mountain Creek, Ala. (1880)	38 in.	35 lb.	3 mls.	382 ft.	224 ft.	5 cars	7,000 lb.	empty	17 tons	Has hauled 7 cars. Grade averages 75 feet per mile for 4 miles. Sharp curves. Round trip of 12 miles each hour. Has taken 4 cars up 4 miles in 20 minutes. No repairs except brake shoes for 8 years. Piston rings never removed.
9½ x 14	21	Emery & Reading........ Dent's Run, Pa. (1888)	56½ in.	35 lb.	6½ mls.	....	250 ft.	2 cars	20,000 lb.	empty	20 tons	
		Same on another part of road	....				75 ft. to 130 ft.	5 cars	20,000 lb.	empty	50 tons	
*9 x 10	38	D. R. Wadley & Co........ Brentwood, Ga. (1881)	60 in.	30 lb.	13 mls.	750 ft.	50 ft.	6 cars	8,000 lb.	12,000 lb.	60 tons	Hauled 10 cars—90 tons. Ran 1 mile with engine only in 1½ minutes. 52 miles, ⅛ cord wood fuel daily.
*9 x 16	21	Whitney & Stinchfield........ Edmore, Mich. (1881)	56¾ in.	30 lb.	7½ mls.	....	52 ft.	12 cars	2,500 lb.	12,500 lb.	90 tons	Hauled 254,475 logs scaling 57,000,000 feet, running 24,940 miles (exclusive of switching, which would add about 25 per cent.) burning 1,263 cords wood fuel in 14 months. Largest haul in 1 month, 5,436,300 feet. Has hauled about 170 tons, and has run 1 mile in 1½ minutes.
*9 x 16	38	W. F. Bailey & Co .... Schlatterville, Ga. (1881)	60 in.	strap	4 mls.	2,500 ft.	100 ft.	4 cars	10,000 lb.	15,000 lb.	50 tons	Curve comes on grade. Can get out 200,000 feet of logs daily.
*9 x 16	21	Buckley & Douglas R. R..... Manistee, Mich. (1881)	36 in.	30 lb.	7 mls.	200 ft.	150 ft.	10 cars	7,100 lb.	empty	35 tons	
*9 x 16	21	Evart & Osceola R. R........ Evart, Mich. (1880)	56½ in.	35 lb.	12 mls.	286 ft.	55 ft.	14 cars	8,000 lb.	16,500 lb.	144 tons	00 miles, 1 cord wood fuel per day of 12 hours. Ran 35 miles per hour with empty train, and made 400 miles in 24 hours, running night and day.
		Same with empty train........	....				190 ft.	14 cars	8,600 lb.	empty	25 tons	

* Change in owner, or location, or service, since report was made.

Size of cylinders.	Page showing style.	Owner and Location, and Date of Report.	Gauge of track in inches.	Weight of rail in pounds per yard.	Length of road in miles.	Radius of sharpest curve in feet.	Grade in feet per mile.	Number of cars hauled at one time.	Weight of each car in pounds.	Load on each car in pounds.	Weight of train in tons of 2,000 lb.	REMARKS.
10 x 16	21	Chippewa River & Menominee R. R....Chippewa Falls, Wis. (1884)	56½ in.	35 lb	6 mls.	375 ft.	26 ft.	20 cars	5,600 lb.	10,000 lb.	156 tons	Grade ¼ mile long. Has hauled 31 cars weighing 242 tons. Usual speed 21 miles per hour. 182 miles, 2 cords wood fuel, 6 tanks water per day of 11 hours.
*10 x 16	21	Blodgett & Byrne....Roscommon, Mich. (1884)	36 in.	30 lb.	7 mls.			15 cars	5,000 lb.	12,000 lb.	127 tons	Grade not measured, but probably over 50 feet per mile. Daily mileage about 182 miles, getting out 150,000 feet of logs daily.
10 x 16	21	Cummer Lumber Co....Cadillac, Mich. (1887)	36 in.	30 lb.	14 mls.		53 ft.	18 cars	4,900 lb.	16,100 lb.	144½ tons	Has hauled 22 cars= 231 tons. Speed 14 to 16 miles per hour. 56 miles, burning 1,350 lbs. coal fuel, and using 3 tanks of water in 10 hours.
*10 x 16	21	Roscommon Lumber Co....Meredith, Mich. (1884)	56½ in.	30 lb.	11 mls.	716 ft.	53 ft.	14 cars	5,000 lb.	12,000 lb.	119 tons	110 to 132 miles. 6,000 lbs. coal fuel, 10 tanks of water per day of 12 hours. Can get out 300,000 feet of logs daily. Have run 262 miles per 24 hours, running night and day.
*10 x 16	21	P. H. Ketcham....Averill, Mich. (1882)	36 in.	30 lb.	21 mls.		53 ft.	20 cars	5,000 lb.	14,000 lb.	190 tons	Takes a run at the grade, which is 800 feet long. Hauled train of 25 cars carrying 102,000 feet, total weight 572 tons, 8 miles, including 53 feet grade, in 8 minutes. 75 to 125 miles during daylight, and 125 to 175

Cylinder	No.	Owner, location (year)	Diam.	Rail	Length			Cars			Capacity	Remarks
												miles in 24 hours. Coal fuel at $4.50 per ton cheaper than wood cut alongside of track.
10 x 16	21	Village Mills Co. Village Mills, Tex. (1888)	36 in.	35 lb.	6 mls.	716 ft.	60 ft.	8 cars	4,000 lb.	20,000 lb.	100 tons	Has hauled 12 cars, 50 miles, hauling 60,000 feet of logs, and could haul 80,000 feet per day of 10 hours easily. Burns refuse trimmings.
10 x 16	23	John Sweet. Rodney, Mich. (1884)	30 in.	25 lb.	12 mls.	477 ft.	85 ft.	15 cars	5,000 lb.	8,000 lb.	97 tons	Has hauled 20 cars weighing 180 tons. Usual speed 15 and best 18 miles per hour. Hauled 15 loaded cars 9 miles, unloaded cars, took water, and brought back empty train in 1 hour. 100 to 126 miles per day of 11 to 12 hours. Grade 1,200 feet long.
*10 x 16	8	Thomas Nester. Ogemaw, Mich. (1879)	56½ in.	40 lb.	7¾ mls.	......	132 ft.	15 cars	4,500 lb.	empty	34 tons	90 miles, 1¼ cords wood fuel daily. Has run 25 miles per hour with loaded train.
10 x 16	21	C. J. L. Meyer. Hermansville, Mich. (1884)	56½ in.	30 lb.	8 mls.	478 ft.	103 ft.	8 cars	9,000 lb.	17,500 lb.	106 tons	Hauled 10 cars weighing 132 tons with reverse lever set in third notch. Grade 700 feet long. 48 to 8½ miles, 1½ to 3 cords wood fuel, 2 to 3 tanks water per day of 10 to 12 hours, getting out 60,000 feet of logs daily. Usual speed 8, and best 25 miles per hour. Full capacity not tested.
†10 x 16	24	Towle Bros. Dutch Flat, Cal. (1878)	30 in.	45 lb.	20 mls.	110 ft.	192 ft.	3 cars	9,500 lb.	23,000 lb.	47 tons	Curve is on grade. Full capacity not tested.
10 x 10	21	Henry, Bayard & Co. Rolfe, Pa. (1865)	56½ in.	50 lb.	6 mls.	477 ft.	211 ft.	4 cars	16,000 lb.	15,000 lb.	62 tons	Has hauled 8 cars—about 100 feet long. Grade about 300 feet long, approached by 211 feet descending grade.

* Change in owner, or location, or service, since report was made.

† This locomotive is also used for hoisting logs from the bottom of a mountain canyon up an incline 1,200 feet long, with a rise of about 650 feet. The engine is run on friction rollers, which are geared to a drum carrying a steel wire rope. The usual load hoisted is 25,000 pounds.

(145)

Size of cylinders.	Page showing style.	Owner and Location, and Date of Report.	Gauge of track in inches.	Weight of rail in pounds per yard.	Length of road in miles.	Radius of sharpest curve in feet.	Grade in feet per mile.	Number of cars hauled at one time.	Weight of each car in pounds.	Load on each car in pounds.	Weight of train in tons of 2,000 lb.	REMARKS.
10 x 16	12	Gurdon Lumber Co., Gurdon, Arks. (1888)	36 in.	35 lb.	4 mls.	382 ft.	231 ft.	8 cars	about 5,000 lb.	empty	20 tons	48 to 64 miles, burning ¾ cord wood fuel, and using 2 tanks water per day of 11 hours. Grade rises 57 feet in 1,300 feet. About $40 repairs in 3 years.
*12 x 18	16	Detroit, Bay City & Alpena R. R., East Tawas, Mich. (1885)	38 in.	48 lb.	50 mls.	212 ft.	26 ft.	45 cars	5,000 lb.	12,000 lb.	383 tons	Has hauled 68 cars=about 516 tons. 4 to 5 round trips of 10 miles daily. Grades slight in either direction, not exceeding 26 feet per mile, and ruling grades in favor of loads.
*12 x 18	16	Wright, Wells & Stone, Ogemaw, Mich. (1884)	56½ in.	35 lb.	25 mls.	955 ft.	50 ft.	20 cars	5,000 lb.	16,000 lb.	210 tons	144 miles, 8 cords of 18-inch wood fuel daily. Grade ¾ mile long.
12 x 18	16	Same on return trip. Geo. W. Pratt, Pelican, Wis. (1888)	56½ in.	35 lb.	16 mls.	716 ft.	157 ft. 80 ft.	20 cars 13 cars	5,000 lb. 5,000 lb.	empty 15,000 lb.	50 tons 180 tons	120 miles, burning 3 cords 4-foot wood fuel, and using 3 tanks of water per day of 14 hours. Grade 2,300 feet long.
12 x 18	21	Puget Mill Co., Port Gamble, Wash. Ter. (1888)	56½ in.	45 lb.	6 mls.	287 ft.	105 ft.	6 cars	4,500 lb.	12,000 lb.	66 tons	58 to 65 miles, burning 1½ cords wood fuel, using 3 tanks of water, and hauling 80,000 feet of logs per day of 12 hours. Logs run up to 5 feet diameter and 100 feet length. 15 miles per hour usual speed.

12 x 18	12	Delta Lumber Co........ Thompson, Mich. (1865)	56½ in.	30 lb.	5¾ mls.	290 ft.	162 ft.	9 cars	3,000 lb.	6,000 lb.	40 tons	About 70 miles. 2½ cords in summer and 4 cords in winter, pine slabs fuel. Power never tested. Gets out about 170,000 feet of logs daily, or about 25,000,000 in one year.
12 x 18	12	Smith & Marbury........ Bozeman, Ala. (1885)	38 in.	35 lb.	7 mls.	382 ft.	200 ft.	8 cars	5,000 lb.	empty	90 tons	Has short grade 204 feet per mile. Grade and curves come together. 42 to 50 miles, burning ⅞ cord of wood fuel, and using 2 tanks of water per day of 10 hours.
12 x 18	16	Loxley & Martin........ Lake Charles, La. (1889)	36 in.	30 lb.	0 mls.	87 ft.	227 ft.	10 cars	4,000 lb.	18,000 lb.	110 tons	Also hauls 12 cars—182 tons. Curve of 256 ft. radius on grade which is ¼ mile long. 54 to 90 miles, 2 cords wood fuel, 2 tanks water per day of 11 hours. Engine rigid wheel base 8 ft. 0 in.
*14 x 20	23	Muscogee Lumber Co...... Pensacola, Fla. (1884)	60 in.	50 lb.	20 mls.	935 ft.	45 ft.	20 cars	17,000 lb.	28,000 lb.	450 tons	60 miles, 3 cords wood fuel, 3 tanks water per day of 11 hours.
14 x 20	16	Chippewa River & Menominee R. R...... Chippewa Falls, Wis. (1884)	56½ in.	35 lb.	8½ mls.	710 ft.	60 ft.	18 cars	5,000 lb.	10,000 lb.	100 tons	Has hauled 22 cars weighing 172 tons. 119 to 153 miles, 3 cords wood fuel, 3 tanks water per day of 11 hours. Usual speed 18, and best 25 miles per hour.
14 x 20	16	Satsop Railroad........ Shelton, Wash. Ter. (1888)	56½ in.	35 lb.	12 mls.	477 ft.	131 ft.	9 cars	5,000 lb.	12,000 lb.	70 tons	100 to 108 miles per day of 12 to 16 hours. 80 miles per hour speed.

* Change in owner, or location, or service, nee report was made.

INDEX.